MEASURING STRESS

A Guide for Health and Social Scientists

EDITED BY

SHELDON COHEN
Carnegie Mellon University

RONALD C. KESSLER
University of Michigan

LYNN UNDERWOOD GORDON
Fetzer Institute

A project of the Fetzer Institute

New York   Oxford   OXFORD UNIVERSITY PRESS   1995

Oxford University Press

Oxford   New York   Toronto
Delhi   Bombay   Calcutta   Madras   Karachi
Kuala Lumpur   Singapore   Hong Kong   Tokyo
Nairobi   Dar es Salaam   Cape Town
Melbourne   Auckland   Madrid

and associated companies in
Berlin   Ibadan

Library of Congress Cataloging-in-Publication Data
Measuring stress : a guide for health and social scientists / edited
by Sheldon Cohen, Ronald C. Kessler, Lynn Underwood Gordon.
     p. cm.   Includes bibliographical references and indexes.
ISBN 0-19-508641-4 (alk. paper)
1. Stress (Psychology)—Measurement.
2. Stress (Physiology)—Measurement.
3. Medicine, Psychosomatic—Research—Methodology.
I. Cohen, Sheldon, 1947–   .
II. Kessler, Ronald C.
III. Gordon, Lynn Underwood.
[DNLM: 1. Stress. Psychological, WM 172 M484 1995]
RC49.M39 1995   616.9—dc20   DNLM/DLC
for Library of Congress   94-13307

2 4 6 8 9 7 5 3 1
Printed in the United States of America
on acid-free paper

Preface

In the winter of 1991 Dr. Lynn Underwood Gordon of the Fetzer Institute invited a small interdisciplinary group of scientists working in the area of psychosocial factors in disease and clinical epidemiology to a meeting in Chicago. The purpose of the meeting was to discuss activities the Fetzer Institute might sponsor that would contribute to improving the methodological rigor and overall quality of research examining the connections between mind and body.

Issues of measurement have been a challenge for clinical medicine, and those interested in assessing the role of psychosocial factors in health and disease face similar challenges. Precise and valid measurement is essential to the construction of meaningful studies. One product of the Chicago meeting was the suggestion that state-of-the-art techniques for measuring various psychosocial factors were slow to disseminate to the wide interdisciplinary group of investigators working in this area. Particular psychosocial measures are often chosen because of their availability or visibility rather than because they are the most appropriate for answering investigators' specific questions. Although several psychosocial factors thought to influence disease processes were discussed at this meeting, including stress, social support, personality, and coping, there was a consensus that the area of stress was one in which there had been important advances in measurement that had not been adequately disseminated, especially across disciplinary lines.

There was also agreement that there is a need for stress concepts to be translated into measurement operations that are accessible to an interdisciplinary group of clinicians and scientists. In popular culture, "stress reduction" has become a goal and is being capitalized upon in a variety of ways, springing from some vague notions of the concept of stress. Many studies indicate that stress can adversely affect our health. Consequently, clarity in the definition of stress and in the expression of that definition in measurement is crucial to designing appropriate and effective stress-reducing interventions.

In response to the Chicago meeting, the Fetzer Institute invited the three of us to edit a volume that would provide a broad range of information about stress measurement to both neophyte and seasoned researchers in the area of stress and disease. We decided that the volume should reflect the different interdisciplinary approaches to stress, help to integrate these approaches, and hence provide a product that would be of use to social and medical scientists alike. We felt that the fact that the three editors represent three of the primary disciplines, psychology, sociology, and epidemiology, would enable us to create a volume with an integrated and balanced view. We proceeded to invite a group of internationally recognized authors from a range of disciplines to contribute to the volume. In our invitations, we emphasized that we wanted to produce a textlike resource for persons from a range of disciplines, with a range of expertise, and hence that these chapters would be harder to write than most. With the support of the Fetzer Institute, we were able to meet twice with our chapter authors, both before and after writing drafts of their chapters, and bring in a number of consultants to provide feedback on chapters at the second meeting. These procedures helped us to guarantee consistency and quality across the volume and encourage conceptual coherence.

The purpose of the volume is to serve as a *resource for state-of-the-art assessment of stress* in studies of physical and psychiatric illness in humans. We expect that the book will be used primarily as a resource for persons conducting research in this area but also in graduate courses in *psychology, psychiatry, sociology, social work, nursing, and epidemiology*. The book includes discussions of how stress is conceptualized, the pathways through which stressors might influence the onset and progression of psychiatric and physical illness, the various methods of measuring stress, and how one decides on appropriate measurement.

Our major goal in producing the volume was to aid researchers in making decisions about the appropriate measures to use in specific studies. Each chapter provides a conceptual underpinning of the approach it addresses, discusses the important measures within the approach, the kinds of studies each is appropriate for, and the various costs and benefits of using each alternative measure. Our intent is to provide valuable information for audiences with a wide range of expertise: to aid persons without extensive experience but at the same time to provide sufficient information for experts to select state-of-the-art measurement instruments.

We are indebted to the Fetzer Institute for its generous support of this endeavor. We express our gratitude to the contributors to this volume, who persevered in the face of editorial onslaught and produced exceptional products in a relatively constrained time frame. Research Scientist Development Awards from the National Institute of Mental Health supported Sheldon Cohen's (MH00721) and Ron Kessler's (MH00507) participation. We thank James House and Alvan Feinstein for their comments during the planning process. We are also indebted to Tracy Herbert, Stan Kasl, Stephen Manuck, and Richard Schulz who served as reviewers, providing exceptional feedback, and who stimulated exciting and challenging discussions about the conceptualization and measurement of stress. We sincerely appreciate their help and note that it is we and not they who are responsible for any errors or misinterpretations. We also want to thank the members of our staffs who have worked so hard in putting this book together—Ruth Dobbins in Kalamazoo, Janet Schlarb and Susan Kravitz in Pittsburgh, and Rose Myers in Ann Arbor—as well as our editor Joan Bossert and the staff of Oxford University Press for their support and professionalism.

Pittsburgh, PA   S.C.
Ann Arbor, MI   R.C.K.
Kalamazoo, MI   L.U.G.

Part IV The Biological Perspective

Contents

editors represent three of the primary disciplines, psychology, sociology, and epidemiology, would enable us to create a volume with an integrated and balanced view. We proceeded to invite a group of internationally recognized authors from a range of disciplines to contribute to the volume. In our invitations, we emphasized that we wanted to produce a textlike resource for persons from a range of disciplines, with a range of expertise, and hence that these chapters would be harder to write than most. With the support of the Fetzer Institute, we were able to meet twice with our chapter authors, both before and after writing drafts of their chapters, and bring in a number of consultants to provide feedback on chapters at the second meeting. These procedures helped us to guarantee consistency and quality across the volume and encourage conceptual coherence.

The purpose of the volume is to serve as a *resource for state-of-the-art assessment of stress* in studies of physical and psychiatric illness in humans. We expect that the book will be used primarily as a resource for persons conducting research in this area but also in graduate courses in *psychology, psychiatry, sociology, social work, nursing, and epidemiology*. The book includes discussions of how stress is conceptualized, the pathways through which stressors might influence the onset and progression of psychiatric and physical illness, the various methods of measuring stress, and how one decides on appropriate measurement.

Our major goal in producing the volume was to aid researchers in making decisions about the appropriate measures to use in specific studies. Each chapter provides a conceptual underpinning of the approach it addresses, discusses the important measures within the approach, the kinds of studies each is appropriate for, and the various costs and benefits of using each alternative measure. Our intent is to provide valuable information for audiences with a wide range of expertise: to aid persons without extensive experience but at the same time to provide sufficient information for experts to select state-of-the-art measurement instruments.

We are indebted to the Fetzer Institute for its generous support of this endeavor. We express our gratitude to the contributors to this volume, who persevered in the face of editorial onslaught and produced exceptional products in a relatively constrained time frame. Research Scientist Development Awards from the National Institute of Mental Health supported Sheldon Cohen's (MH00721) and Ron Kessler's (MH00507) participation. We thank James House and Alvan Feinstein for their comments during the planning process. We are also indebted to Tracy Herbert, Stan Kasl, Stephen Manuck, and Richard Schulz who served as reviewers, providing exceptional feedback, and who stimulated exciting and challenging discussions about the conceptualization and measurement of stress. We sincerely appreciate their help and note that it is we and not they who are responsible for any errors or misinterpretations. We also want to thank the members of our staffs who have worked so hard in putting this book together—Ruth Dobbins in Kalamazoo, Janet Schlarb and Susan Kravitz in Pittsburgh, and Rose Myers in Ann Arbor—as well as our editor Joan Bossert and the staff of Oxford University Press for their support and professionalism.

Pittsburgh, PA   S.C.
Ann Arbor, MI   R.C.K.
Kalamazoo, MI   L.U.G.

Preface

In the winter of 1991 Dr. Lynn Underwood Gordon of the Fetzer Institute invited a small interdisciplinary group of scientists working in the area of psychosocial factors in disease and clinical epidemiology to a meeting in Chicago. The purpose of the meeting was to discuss activities the Fetzer Institute might sponsor that would contribute to improving the methodological rigor and overall quality of research examining the connections between mind and body.

Issues of measurement have been a challenge for clinical medicine, and those interested in assessing the role of psychosocial factors in health and disease face similar challenges. Precise and valid measurement is essential to the construction of meaningful studies. One product of the Chicago meeting was the suggestion that state-of-the-art techniques for measuring various psychosocial factors were slow to disseminate to the wide interdisciplinary group of investigators working in this area. Particular psychosocial measures are often chosen because of their availability or visibility rather than because they are the most appropriate for answering investigators' specific questions. Although several psychosocial factors thought to influence disease processes were discussed at this meeting, including stress, social support, personality, and coping, there was a consensus that the area of stress was one in which there had been important advances in measurement that had not been adequately disseminated, especially across disciplinary lines.

There was also agreement that there is a need for stress concepts to be translated into measurement operations that are accessible to an interdisciplinary group of clinicians and scientists. In popular culture, "stress reduction" has become a goal and is being capitalized upon in a variety of ways, springing from some vague notions of the concept of stress. Many studies indicate that stress can adversely affect our health. Consequently, clarity in the definition of stress and in the expression of that definition in measurement is crucial to designing appropriate and effective stress-reducing interventions.

In response to the Chicago meeting, the Fetzer Institute invited the three of us to edit a volume that would provide a broad range of information about stress measurement to both neophyte and seasoned researchers in the area of stress and disease. We decided that the volume should reflect the different interdisciplinary approaches to stress, help to integrate these approaches, and hence provide a product that would be of use to social and medical scientists alike. We felt that the fact that the three

Contributors

ANDREW BAUM
Department of Psychiatry and Pittsburgh Cancer Institute, University of Pittsburgh, Pittsburgh, PA 15213.

NIALL BOLGER
Department of Psychology, New York University, New York, NY 10003.

GEORGE W. BROWN
Department of Social Policy and Social Science, Bedford College, University of London, London, Great Britain WC1B 3RA.

SHELDON COHEN
Department of Psychology, Carnegie Mellon University, Pittsburgh, PA 15213.

JOHN ECKENRODE
Human Development and Family Studies, Cornell University, Ithaca, NY 14853.

JENNIFER J. FALCONER
Department of Medical Psychology, Uniformed Services University of the Health Sciences, Bethesda, MD 20814.

RONALD GLASER
Department of Medical Microbiology and Immunology, Brain, Behavior, Immunity and Health Program, College of Medicine, Ohio State University, Columbus, OH 43210.

LYNN UNDERWOOD GORDON
Fetzer Institute, 9292 West KL Avenue, Kalamazoo, MI 49009.

NEIL GRUNBERG
Department of Medical Psychology, Uniformed Services University of the Health Sciences, Bethesda, MD 20814.

JOHN M. KELLEY
Department of Psychology, University of Oregon, Eugene, OR 97403.

RONALD C. KESSLER
Department of Sociology and Institute of Social Research, University of Michigan, Ann Arbor, MI 48016.

JANICE K. KIECOLT-GLASER
Department of Psychiatry, Brain, Behavior, Immunity and Health Program, College of Medicine, Ohio State University, Columbus, OH 43210.

DAVID S. KRANTZ
Department of Medical Psychology, Uniformed Services University of the Health Sciences, Bethesda, MD 20814.

STEPHEN J. LEPORE
Department of Psychology, Carnegie Mellon University, Pittsburgh, PA 15213.

SCOTT M. MONROE
Department of Psychology, University of Oregon, Eugene, OR 97403.

ARTHUR A. STONE
Department of Psychiatry, SUNY at Stony Brook, Stony Brook, NY 11794.

R. JAY TURNER
Department of Sociology, University of Toronto, Toronto, Ontario, Canada M5T 1P9.

ELAINE WETHINGTON
Department of Human Development and Family Studies, Cornell University, Ithaca, NY 14853.

BLAIR WHEATON
Department of Sociology, University of Toronto, Toronto, Ontario, Canada M5T 1P9.

MEASURING
STRESS

PART I

Conceptualizing Stress and Its Relation to Disease

This part addresses alternative definitions of stress, pathways through which stress could influence psychiatric and physical disorders, and issues involved in choosing between various levels of stress measurement. We discuss disciplinary differences in the definition and measurement of stress and suggest a unifying model relating various approaches to the study of disease. In particular, we address the roles of environmental, psychological, and biological conceptualizations of stress as contributors to disease. Possible typologies of stress based on temporal characteristics—for example, acute versus chronic stress—are also discussed. Chapter 1 is designed to help investigators formulate questions and begin to choose possible measurement options. The remainder of the volume provides detailed information for choosing specific measures.

1

Strategies for Measuring Stress in Studies of Psychiatric and Physical Disorders

Sheldon Cohen, Ronald C. Kessler, and Lynn Underwood Gordon

Stress has long been a major focus among researchers interested in environmental and psychosocial influences on health. However, the way in which the term "stress" has been used in this voluminous literature has not been consistent. Indeed, some commentators have gone as far as to argue that the term stress has so many different meanings that it has become a useless concept (Ader, 1980; Elliot & Eisdorfer, 1982). Although we disagree with this premise, we recognize that there is confusion about the meaning and measurement of stress. This volume attempts to clear up some of that confusion by providing conceptual and practical advice for formulating questions about the relation between stress and disease and for selecting appropriate stress measures to test these questions.

What Is Stress?

As noted above, there is disagreement about the meaning of the term "stress." Numerous definitions have been provided, varying in the extent to which they emphasize stressful events, responses, or individual appraisals of situations as the central characteristic of stress (e.g., Appley & Trumbull, 1967; Mason, 1975; McGrath, 1970). We have no illusions that we can resolve the differences among these perspectives in this volume. However, we do see a strong commonality among these approaches that allows them to be integrated in a theoretical model of the role of stress in disease. They all share an interest in a process in which *environmental demands tax or exceed the adaptive capacity of an organism, resulting in psychological and biological changes that may place persons at risk for disease.*

Three broad traditions of assessing the role of stress in disease risk can be distinguished. The environmental tradition focuses on assessment of environmental events or experiences that are normatively (*objectively*) associated with substantial adaptive demands. The psychological tradition focuses on individuals' *subjective* evaluations of their abilities to cope with the demands posed by specific events or experiences. Finally, the biological tradition focuses on activation of specific physiological systems that have been repeatedly shown to be modulated by both psychologically and physically demanding conditions. The following sections provide a brief orientation to these traditions and attempt to distill the central assumptions of each. We then present an organizational model of the stress process that shows how these traditions relate to one another. Our premise is that each tradition focuses on a different stage of the process through which environmental demands are translated into psychological and biological changes that place people at risk for disease. Finally, we discuss the conceptual and practical issues involved in selecting appropriate categories of stress measurement. Once appropriate categories are selected, subsequent chapters of this volume can be consulted to help select measures within each category.

Throughout the volume we use the term stress exclusively to refer to the general process through which environmental demands result in outcomes deleterious to health. However, we distinguish between the components of the process by referring to environmental experiences as *environmental demands, stressors, or events;* to subjective evaluations of the stressfulness of a situation as *appraisals or perceptions of stress;* and to affective, behavioral, or biological responses to stressors or appraisals as *stress responses.*

The Environmental Stress Perspective

Most evidence on the role of stressors in human disease has derived from interest in stressful life events. Interest in the role of life events in illness began with the work of Adolf Meyer in the 1930s. Meyer advocated that physicians fill out a life chart as part of their medical examination of ill patients (Lief, 1948; Meyer, 1951). Meyer believed that the life events elicited in this way could be shown to have etiologic importance for a variety of physical illnesses. Meyer's ideas were highly influential and led to a substantial body of research which, by the late 1940s, had documented that stressful life events were associated with a variety of physical illnesses (see review by Wolff, Wolf, & Hare, 1950). Although some of this early work was based on inadequate research designs, a number of studies were quite impressive. The work of Wolff and his associates, for example, followed a large sample of telephone operators over many years and documented that illness was much more likely to occur during periods of inordinate demands, frustrations, and losses than at other times (Hinkle & Wolff, 1958).

An important advance in this area of research came in 1957 when Hawkins and his collaborators developed the Schedule of Recent Experiences (SRE) in an effort to systematize Meyer's life chart (Hawkins, Davies, & Holmes, 1957). This instrument was used by a great many researchers over the next decade to document associations between stressful life events and heart disease, skin disease, and many others (reviewed by Holmes & Masuda, 1974). In a subsequent modification of the SRE, the Social Readjustment Rating Scale (SRRS), each event was assigned a standardized weight based on judges' ratings of the degree of difficulty required to adjust to the event (Holmes & Masuda, 1974). These weights were called "life change units" (LCU). The summing of LCUs associated with reported events allowed for a summary measure of environmental stressors (Holmes & Rahe, 1967). This instrument had an enormous impact on research on the relations between life events and illness, due in large part to the documentation of dramatic associations, such as an effect of stressful life events on sudden cardiac death (Rahe & Lind, 1971). It also had an important conceptual impact on the field in advancing the notion of the life change unit and the conceptual model underlying the creation of this metric, which argued that the effects of stressors operate largely through the creation of excessive adaptive demands. This conception led users of the SRRS to be more concerned with the magnitude of life change than with whether the change was positive (e.g., a promotion) or negative (e.g., a job loss).

Beginning in the 1970s, a new generation of stressful life event researchers began to challenge many of the basic assumptions involved in the construction and scoring of the SRRS. New ideas were advanced about the implications of different means of weighting and summing multiple events into cumulative scales (Shrout, 1981). A subjective element was introduced into some modifications of the SRRS by having individuals estimate the stressfulness of their own experiences as a way of generating measures that are more sensitive indicators of event stressfulness than judges' ratings (e.g., Sarason, Johnson, & Siegel, 1978). More drastic differences are reflected in the development of a life event interview in which investigators rate the importance of events while taking into account the context in which they occur (Brown & Harris, 1978). The investigator-based rating is an attempt to estimate the impact of an event in a specific context for the average person, avoiding individual subjective reactions. A major distinction among competing measures of life events in the current literature is between these contextual measures and more traditional checklist measures (Brown & Harris, 1989; Dohrenwend, Raphael, Schwartz, Stueve, & Skodol, 1993). Separate chapters on these two approaches are included in Section II of this volume.

New concerns were also raised during this period that existing life event scales may not include an adequate and representative sample of the major events that occur in people's lives. Newer checklists were developed to expand the range of experiences evaluated (Dohrenwend, Askenasy, Krasnoff, & Dohrenwend, 1978). Scales were also developed to assess stressful events in specific populations whose experiences might be different from those represented on the more general SRRS. These included scales for children (e.g., Sandler & Ramsay, 1980), adolescents (e.g., Newcomb, Huba, & Bentler, 1981) and the elderly (e.g., Murrell, Norris, & Hutchins, 1984).

The basic assumptions of the SRRS, that the effects of all stressful events are cumulative and that the change per se is the most important dimension of stressors, were also challenged during the same time period (Paykel, 1974). Some newer life-event scales were based on a multidimensional conception of stressors that separately assessed the extent of threat, loss, danger, and other aspects of stressful events (Brown & Harris, 1978).

On the substantive side, there has been a continuation of basic research to document the effects of stressful events on a variety of physical and mental health outcomes using newer stressful life event measures (Chapters 2 & 3, this volume).

There also has been interest in studying the cumulative effects of experiencing two or more stressful life events in the same short interval of time (McGonagle & Kessler, 1990) and the joint effects of experiencing a stressful event in the context of an ongoing chronic stressor in the same life domain (Wheaton, 1990).

In addition, there is now considerable interest in studying *vulnerability factors*—characteristics that make people more or less susceptible to stressor-induced disease. This interest derives from the repeated finding that although environmental stressors are often associated with illness onset, the majority of people confronted with extreme or traumatic stressful events do not become ill (Thoits, 1983). Differential vulnerability to the health-damaging effects of environmental stressors has been documented in a number of investigations (see review by Kessler, Price, & Wortman, 1985). A search for the determinants of this differential vulnerability has become the main focus of researchers interested in the role of life events in illness. In addition, the decade of the 1980s saw a movement away from an earlier tradition of focusing exclusively on the acute health-damaging effects of discrete life events toward an investigation of the long-term health-damaging effects of chronic stressors (Chapter 5, this volume). Work stressors (Neilson, Brown, & Marmot, 1989), marital disharmony (Beach, Sandeen, & O'Leary, 1990), and work–family conflicts (Eckenrode & Gore, 1990) have been the primary areas of investigation in research on chronic stressor effects. In addition, there is a new interest in the cumulative effects of minor daily stressors on both emotional health (Bolger, DeLongis, Kessler, & Schilling, 1989) and physical health (Chapter 4, this volume; Stone, Reed, & Neal, 1987). In all of this work, the researcher attempts to identify characteristics of the environment that promote illness. In the initial stages of this work the focus is on description. As the work evolves, the focus shifts to the processes involved in creating the observed association, the topic of Parts III and IV of this volume.

Finally, although stressful events have been studied primarily as risk factors for disease, it is becoming increasingly clear that confronting and adapting to stressful events can result in positive outcomes such as personal growth, reprioritization of life goals, increased feelings of self-esteem and self-efficacy, and strengthening of social networks. A greater emphasis on the benefits of stressful events for successful adaptors is likely in the future and would broaden our understanding of the stress process.

The Psychological Stress Perspective

The psychological stress tradition places emphasis on the *organism's perception and evaluation* of the potential harm posed by objective environmental experiences. When their environmental demands are perceived to exceed their abilities to cope, individuals label themselves as stressed and experience a concomitant negative emotional response. Psychological models of stress argue that events influence only those persons who appraise them as stressful—that is, perceive stress. It is important to emphasize that stress appraisals are determined not solely by the stimulus condition or the response variables, but rather by persons' interpretations of their relationships to their environments. That is, the perception that one is experiencing stress is a product of both the interpretation of the meaning of an event and the evaluation of the adequacy of coping resources.

The most influential model of the appraisal process has been the one proposed by Lazarus (Lazarus & Folkman, 1984). In the original formulation of his model, Lazarus (1966) argued that an appraisal of a stimulus as threatening or benign, termed *primary appraisal,* occurs between stimulus presentation and stress reaction. In his later writings, Lazarus (1977, 1981) argued that a situation will also result in a stress reaction if it is evaluated as a harm/loss, threat, or challenge. Primary appraisal is presumed to depend on two classes of antecedent conditions: the perceived features of the stimulus situation and the psychological structure of the individual. Some stimulus factors affecting primary appraisal include the imminence of harmful confrontation, the magnitude or intensity of the stimulus, the duration of the stimulus, and the potential controllability of the stimulus. Factors within individuals that affect primary appraisal include their beliefs about themselves and the environment, the pattern and strength of their values and commitments, and related personality dispositions. When a stimulus is appraised as requiring a coping response, individuals evaluate their resources in order to determine whether they can cope with the situation—that is, eliminate or at least lessen the effects of a stressful stimulus. This process is termed "secondary appraisal." Coping responses may involve actions designed to directly alter the threatening conditions (e.g., fight or flight) or thoughts or actions whose goals are to relieve the emotional stress response (i.e., body or psychological disturbances). The latter group of responses, referred to as "emotionally focused" coping, may be somatically oriented —for example, the use of tranquilizers, or intrapsychic responses such as denial of danger (Lazarus, 1975). If one perceives that effective coping responses are available, then the threat is short-circuited and no stress response occurs. If, on the other hand, one is uncertain that she or he is capable of coping with a situation that has been appraised as threatening or otherwise demanding, stress is experienced. It is important to note that this process of evaluating the demands of a situation and evaluating one's ability to cope not only occurs at the onset of a stressful event but often recurs during the course of the event (cf. Folkins, 1970; Lazarus, 1981). Thus, an event that is initially appraised as threatening may be later reappraised as benign, and coping strategies that are initially found to be lacking may later be found to be adequate. Conversely, events that one initially evaluates as nonthreatening may be later reevaluated as stressful. Although it is recognized that certain events are almost universally appraised as stressful (e.g., the death of a loved one), the impact of even these events can be expected to depend on an individual's appraisal of the threat entailed and his or her ability to cope with it. For example, the death of a spouse for someone with neither family nor friends may be experienced as more severe than the same event for someone with close ties to family and friends.

As suggested earlier, appraisals of threat elicit negative emotional responses. They also can elicit a range of other outcomes including self-reported annoyance, changes in health practices such as smoking, drinking alcohol, diet, exercise, and sleeping; changes (usually deficits) in performance of complex tasks; and alterations in interpersonal behaviors (Cohen, Evans, Krantz, & Stokols, 1986). Unfortunately, psychological stress models tend to be vague in their predictions of the particular measures that will be affected in any instance, and of the nature of the relations among these outcome measures.

The Biological Stress Perspective

The biological perspective focuses on the activation of physiological systems that are particularly responsive to physical and psychological demands. Prolonged or repeated activation of these systems is thought to place persons at risk for the development of a range of both physical and psychiatric disorders. Two interrelated systems that are viewed as the primary indicators of a stress response are the sympathetic–adrenal medullary system (SAM) and the hypothalamic–pituitary–adrenocortical axis (HPA). Although detailed descriptions of these two systems and their relations to each other are beyond the scope of this chapter (see Baum, Singer, & Baum, 1981; Levi, 1972; and Chapter 8, this volume), each is discussed in brief in order to provide a basic understanding of their roles in the stress process.

Sympathetic–Adrenal Medullary System

Interest in the impact of SAM activation on bodily reactions to emergency situations may be traced to Walter Cannon's early work on the flight or flight response (Cannon, 1932). Cannon proposed that the SAM system reacts to various emergency states with increased secretion of the hormone epinephrine. There is a large body of evidence indicating increased output of epinephrine and norepinephrine in response to a wide variety of psychosocial stressors (Levi, 1972). Other components of the SAM response elicited by stressors include increased blood pressure, heart rate, sweating, and constriction of peripheral blood vessels. It has been claimed that if SAM activation is excessive, is persistent over a period of time, or is repeated too often, it may result in a sequence of responses that culminate in illness. The responses include functional disturbance in various organs and organ systems (cf. Dunbar, 1954) and ultimately permanent structural changes of pathogenic significance at least in predisposed individuals (e.g., Raab, 1971). Particularly culpable in this regard is the secretion of the hormones epinephrine and norepinephrine by the adrenal medulla and/or sympathetic nerve endings. Excessive discharge of these substances is believed to induce many of the pathogenic states associated with the perception of stress including (1) suppression of cellular immune function (e.g., Rabin, Cohen, Ganguli, Lysle, & Cunnick, 1989); (2) hemodynamic effects, such as increased blood pressure and heart rate (McCubbin, Richardson, Obrist, Kizer, & Langer, 1980); (3) provocation of variations in normal heart rhythms (ventricular arrhythmias) believed to lead to sudden death (Herd, 1978); and (4) production of neurochemical imbalances that contribute to the development of psychiatric disorders (Anisman & Zacharko, 1992).

Hypothalamic–Pituitary–Adrenocortical Axis (HPA)

The hormonal responses of the HPA axis were emphasized in Hans Selye's (e.g., 1956, 1974) influential description of a nonspecific (general) physiological reaction that occurs in response to excessive stimulation. Selye argued that pathogens, physical stressors (e.g., shock or noise), and psychosocial stressors all elicit the same pattern of physiological response. This response is said to proceed in a characteristic three-stage pattern referred to as the general adaptation syndrome (GAS). During the first stage of the GAS, the *alarm stage,* the organism's physiological changes reflect the initial reactions necessary to meet the demands made by the stressor agent. The anterior pituitary gland secretes adrenocorticotrophic hormone (ACTH), which then activates the adrenal cortex to secrete additional hormones (corticosteroids [primarily cortisol in humans]). The hormone output from the adrenal cortex increases rapidly during this stage. The second stage, *resistance,* involves a full adaptation to the stressor with consequent improvement or disappearance of symptoms. The output of corticosteroids remains high but stable during the resistance stage. Finally, the third stage, *exhaustion,* occurs if the stressor is sufficiently severe and prolonged to deplete somatic defenses. The anterior pituitary and the adrenal cortex lose their capacity to secrete hormones, and the organism can no longer adapt to the stressor. Symptoms reappear, and, if the stress response continues unabated, vulnerable organs (determined by genetic and environmental factors) will break down. This breakdown results in illness and ultimately death. Selye argued that any noxious agent, physical or psychosocial in nature, would mobilize a similar GAS response. In contrast, critiques of Selye's model suggest that each stressor elicits its own distinct physiogical reactions (Lazarus, 1977; Mason, 1975). These authors agree that there is a nonspecific physiological response to stressors. They argue, however, that the response is a concomitant of the *emotional reaction* that occurs when situations are appraised as stressful. When conditions are designed to reduce the psychological threat that might be engendered by laboratory procedures, there is no nonspecific reaction to a physical stressor (Mason, 1975). For example, by minimizing competitive concerns and avoiding severe exertion, the danger that young men would be threatened by treadmill exercise was reduced, and the GAS pattern was not found. It is noteworthy that, in Selye's later work (1974, 1980) he acknowledged that there are both specific as well as general (nonspecific) factors in physiological response to a stressor but maintained that the nonspecific response is not always psychologically mediated. He also suggested that the GAS does not occur (or is at least not destructive) in response to all kinds of stressors. For example, he suggested that there may be a pleasant stress of fulfillment and victory and a self-destructive distress of failure, frustration, and hatred. However, there is little empirical evidence to support this position.

Since the late 1970s, interest in the biological bases of psychiatric disorders has stimulated an alternative focus on the HPA. Most of this work has pursued the possible role of HPA disregulation in depression. Relatively pronounced HPA activation is common in depression, with episodes of cortisol secretion being more frequent and of longer duration among depressed than among other psychiatric patients and normals (Stokes, 1987). However, it is still unclear whether the hyper-HPA activation is a cause or effect of depressive disorders. HPA regulation may play a role in other psychiatric disorders as well. For example, anxious patients tend to have higher cortisol levels than normal controls (Sachar, 1975).

Other Stress Associated Changes

Although hormones of the SAM and HPA are those most often discussed as the biochemical substances involved in stress responses, alterations in a range of other hormones, neurotransmitters, and brain substances have also been found in response to stress and may play an important role in stress influences on health. These include stressor-associated elevations in growth hormone and prolactin secreted by the pituitary gland, and in the natural opiates beta endorphin and enkephalin re-leased in the brain (see Chapter 8, this volume; also Baum, Grunberg, & Singer, 1982). These substances are also thought to play a role in both immune-mediated (Rabin et al., 1989) and psychiatric diseases (Stokes, 1987).

A Unifying Model of the Stress Process

The perspectives represented by the three traditions discussed earlier can be viewed as emphasizing different points in the process through which objective environmental experiences can influence disease. A model integrating these approaches is presented in Figure 1.1. The sequential relations between the central components of the model (dark arrows) can be described as follows. When confronting environmental demands, people evaluate whether the demands pose a potential threat and whether sufficient adaptive capacities are available to cope with them. If they find the environmental demands taxing or threatening, and at the same time view their coping resources as inadequate, they perceive themselves as under stress. The appraisal of stress is presumed to result in negative emotional states. If extreme, these emotional states may directly contribute to the onset of affective psychiatric disorders. They may also trigger behavioral or physiological responses that put a person under risk for psychiatric and physical illness. This model implies that each sequential component of the stress process is more proximal to and hence more predictive of the illness outcome. For example, a disease-relevant biological stress-response measure should be a better predictor of a disease outcome than measures of stressful life events or perceived stress.

We thought it important that the model also represent the possibility that environmental demands can put persons at risk for disorder even when appraisal does not result in perceptions of stress and negative emotional responses. This is represented by the arrow directly linking environmental demands to physiological or behavioral responses. For example, it has been argued that the process of coping itself (even when it is successful and environmental demands are appraised as benign) may *directly* result in physiological and behavioral changes that place persons at risk for disease (Cohen et al., 1986; Cohen, Tyrrell, & Smith, 1993).

We want to emphasize that this is a *heuristic model* designed to illustrate the

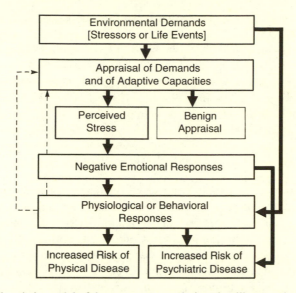

Figure 1.1 A heuristic model of the stress process designed to illustrate the potential integration of the environmental, psychological, and biological approaches to stress measurement.

potential integration of the environmental, psychological, and biological approaches to stress measurement. The model is primarily unidirectional (flowing from environmental demands to disease) and does not include all possible pathways linking these concepts. We identified two of *many possible* feedback loops (dashed lines) to illustrate the potential role of feedback in the model. One of the loops suggests that emotional states may alter appraisals. For example, depressed affect may result in negatively biased views of either the threat posed by stressors or the adequacy of one's own resources. The other loop suggests that physiological arousal may alter appraisals and emotional responses: For example, persons may mistakenly attribute arousal that was elicited by exercise, drugs, or nonrelevant emotional responses to a stressor (Schachter & Singer, 1962). The exclusion of alternative paths is not intended to reflect hypotheses about their existence.

Historically, the environmental, psychological, and biological traditions have each focused on a specific part of the model, thus often ignoring other parts. For example, sociologists and epidemiologists have addressed the question of whether life events increase disease risk but usually ignore the psychological and biological pathways through which this influence might occur. Psychologists have focused on the role of appraisal and emotional response in disease risk, with less emphasis on the environmental causes of these states and the biological pathways responsible for links between psychological states and disease. Finally, biological stress researchers have focused primarily on the links between stressors and hormonal and cardiovascular response, and between these responses and disease risk, without concerning themselves with the psychological pathways through which stressors might influence biological states.

How Could Stress Influence Disease?

We have provided a generic model of the stress process. In this section, we elaborate the implications of the generic model for physical and psychiatric illness. Our hope is that this presentation of how various conceptions of stress can influence illness will help generate appropriate questions and, consequently, the selection of measures and research designs.

Stress and Physical Illness

In general, stressors are thought to influence the pathogenesis of physical disease by causing negative affective states (such as anxiety and depression) which in turn exert direct effects on biological processes or behavioral patterns that influence disease risk (see Cohen et al., 1986; Krantz, Glass, Contrada, & Miller, 1981). The primary biological pathway linking emotions to disease is thought to be hormonal. Hormonal responses associated with stressful experiences include elevations in the catecholamines epinephrine and norepinephrine secreted by the adrenal medulla, in cortisol secreted by the adrenal cortex, in growth hormone and prolactin secreted by the pituitary gland, and in the natural opiates beta endorphin and enkephalin released in the brain (see Baum, Grunberg, & Singer, 1982). A number of these hormones have been implicated in the pathogenesis of cardiovascular disease (Herd, 1986) and diseases involving the immune system including cancer, infectious diseases, and autoimmune diseases (Cohen & Williamson, 1991; Laudenslager, 1988; Rabin et al., 1989). Emotionally induced responses of the cardiovascular system such as increased heart rate and blood pressure have also been implicated in the development of cardiovascular disease and of immune changes that might alter susceptibility to immune-mediated disease (Herbert et al., 1994; Manuck et al., 1992). As discussed earlier, emotional responses are not always required for stressful events to influence disease processes (Cohen, Tyrrell, & Smith, 1993). The effort involved in actively coping with a stressor may also alter many of the same biological processes influenced by the emotional response and hence influence the development of disease independently of the emotional response (Cohen et al., 1986).

Behavioral changes occurring as adaptations or coping responses to stressors may also influence disease risk. For example, persons exposed to stressors or viewing themselves as under stress tend to engage in poor *health practices*. They may smoke more, drink more alcohol, eat poorly, exercise less, and sleep less (e.g., Cohen & Williamson, 1988; Conway, Vickers, Ward, & Rahe, 1981). Smoking, drinking alcohol, and poor diets have been established as risk factors for a range of different physical illnesses. Both stressors and negative affect have also been associated with failure to *comply* with medical regimens. Such failure could result in more severe and longer-lasting illness, either because undesirable behaviors aggravate existing problems or because failure to perform desirable behaviors (e.g., following medication regimens) results in disease progression. Other stressor-elicited behaviors—for example, unsafe sexual practices or poor hygienic practices—could also increase exposure to infectious agents.

Stress and Illness Behaviors

Stress may also influence behaviors that appear to be manifestations of a disease state. Of particular interest is the role of stress in recognizing and acting on symptoms. The recognition and reporting of symptoms, and seeking of medical care often indicate underlying pathology. However, stress and other psychological factors can independently influence these behaviors (Cohen & Williamson, 1991). Because appraised stress often triggers physiological arousal, people under stress may be more (possibly overly) attentive to their internal physical states (Pennebaker, 1982). Stress may also facilitate the labeling of sensations as symptoms because people are reminded (in cognitive parlance, a schema is triggered) of previous times when stress was associated with symptoms or simply because they believe that stress triggers symptoms. Alternatively, stressors or stress appraisals may result in physical sensations whose causes are mistakenly attributed to disease symptoms rather than the stressor (e.g., Mechanic, 1972; Schachter & Singer, 1962). Labeling symptom constellations as disease may similarly be activated by stress–disease schemas. For example, it is widely believed that stressors cause the recurrence of oral herpes. Under stress, a minor oral lesion that would be ignored under nonstressful conditions may be defined as disease recurrence. Reports of symptoms and illness are also ways to avoid stressful situations (Mechanic, 1977). The prototypical example is the child who reports symptoms to avoid attending school on an especially stressful day (playing ill). Finally, stressors or stress appraisals may influence the decision to seek medical care when persons label themselves as ill. The perception of stress could interfere with deciding whether it is necessary to seek care, increasing care-seeking for minor symptoms or decreasing care-seeking for serious ones. Persons under stress may also seek medical care unnecessarily because medical providers are viewed as persons to whom one can confide problems. Stressors could also *decrease* care-seeking because the time demands of many stressors make such visits inconvenient (Schulz, Visintainer, & Williamson, 1990).

Stress and Psychiatric Illness

As noted above, most models of the relations between stress and physical illness postulate an intervening link through negative affective states such as anxiety and depression. But how is it that stressors cause these affective states, particularly those that are so extreme and enduring that they are considered disorders in their own right? There are two broad perspectives on this question. One holds that stressful experiences are sometimes so extreme that they naturally lead to enduring fear or sadness. This is the notion, for example, underlying the diagnosis of post-traumatic stress disorder, a disorder defined by the American Psychiatric Association as one in which there is enduring distress related to recurrent recollections of a traumatic event that is so far "outside the range of usual human experience . . . that (it) would be markedly distressing to almost anyone" who experienced it (American Psychiatric Association, 1987:250). The second perspective holds that stressful events lead to psychiatric disorder only in the presence of some preexisting personal vulnerability. A number of different vulnerability factors have been postulated for different disorders. Some of these are thought to be environmental in origin (U.S. Congress Office of Technology Assessment, 1992). For example, some researchers believe that exposure to abnormal parental attachment behaviors early in life can lead children to develop disturbed interpersonal styles which persist throughout their lives and create vulnerability to depression (Kessler & Magee, 1993). This vulnerability is thought to be triggered by stressful experiences associated with interpersonal loss, such as death of a loved one, divorce, or forced residential relocation. Other vulnerabilities are thought to be biological in origin. For example, there is evidence consistent with the possibly that HPA abnormalities are involved in vulnerability to depression (Anisman & Zacharko, 1992; Checkley, 1992) and that these HPA abnormalities may be under genetic control (Checkley, 1992).

Temporal Characteristics of Stressors, Appraisals, and Stress Responses

One of the most challenging issues in the measurement of stress is the characterization of the temporal course of stressors, appraisals, and stress responses. Most of the discussion of the role of stress duration focuses on the chronicity of stress. Although the terms "acute" and "chronic" stress are used liberally in the literature, there have been few attempts (for an exception, see Brown & Harris, 1989) to set even arbitrary standard cutoffs delimiting acute from chronic durations. Our own view is that such cutoffs should be attempted only in the context of considering the implications for a specific outcome. For example, 3 months of persistent stress might prove important in risk for depression, and prove inconsequential for a disease that develops over years such as coronary artery disease. Assuming we were correct about the implications of these durations, it might be appropriate to define 3 months of persistent stress as "chronic" in the study of depression, but it would be uninformative to call 3 months of stress chronic in the study of heart disease (see Cohen & Matthews, 1987; Cohen, Kaplan, & Manuck, 1994).

A view of temporal characteristics of stress consistent with the process model discussed earlier is proposed by Baum and his colleagues (Baum, Cohen, & Hall, 1993). They categorize stress duration through the use of a 2 × 2 × 2 matrix that crosses duration of event exposure (i.e., event present for short or long duration), duration of perceived threat (i.e., appraised threat or demand present for short or long duration), and duration of stress responding (i.e., behavioral, emotional, or physiological stress responses present for short or long duration). This procedure suggests a more sensitive approach to understanding the role of stress duration. For example, consider the difference between a persistent stressful event that is no longer appraised as stressful or responded to with a stress response, and a stressful event that has terminated but continues to be appraised as stressful and responded to with a stress response (e.g., traumatic experiences).

Other work highlights some of the complexities of this issue by emphasizing the importance of temporal characteristics other than duration such as continuousness or repetitiveness. The following classification of stressful events (from Elliott & Eisdorfer, 1982) provides an indication of the complexity of this issue: (1) *acute time-limited events* (e.g., awaiting surgery); (2) *stressful event sequences*—when one event initiates a series of different events that occur over an extended period of time (e.g., bereavement or being fired from a job); (3) *chronic intermittent stressful events*—events that occur periodically (once a week, once a month, or once a year; e.g., sexual difficulties or conflicts with neighbors); and (4) *chronic stress conditions*—situations that may or may not be initiated by a discrete event (e.g., being disabled, chronic job stress).

The temporal course of events, stress appraisals, and stress responses have multiple implications for understanding the risk for disease. For example, one can ask how long an exposure, appraisal, or response is required to alter a disease process. Does persisting exposure result in a greater impact or in habituation? Are stress responses maintained over long periods, or do response mechanisms fatigue or trigger feedback mechanisms? What roles do repetitiveness and interval between repetitions play in these processes? Also of interest is the role that chronic stressors play in moderating the effect of more acute stressors. Unfortunately, there are few domains in which we know very much about the temporal role of stress. It is hoped that future longitudinal research addressing specific questions about the temporal course of stress will begin to provide us with a better understanding about how these issues should be treated.

Matching Measures, Designs, and Research Questions

The choice of appropriate stress measures depends on the *disease* (or stage of disease) under study, on the *specific question* posed by the investigator about the relation between stress and disease, and on other *methodological and practical issues*. Although we discuss each of these criteria separately, all three criteria need to be optimized if the most appropriate measure is to be selected.

Matching the Temporal Courses of the Stress Measure and Disease

Choosing an appropriate stress measure requires investigators to be informed about the disease outcome they are studying. Of particular importance is an understanding of the *temporal course* of the disease or disease stage under investigation (Cohen, Kaplan, & Manuck, 1994; Cohen & Matthews, 1987). Take, for example, the possible role of stressors in coronary artery disease (CAD). Atherosclerosis (occlusion of the coronary arteries) does not occur suddenly, over a few weeks or months, or even over several years. Instead it develops over decades. What kind of stressor measure would be most appropriate in studying the potential impact of stressors on CAD? Answering this question requires the investigator first to specify plausible pathways through which stressors might influence the development of the disease. Logically, there are two primary ways in which stressors could play a part in a disease with this course of development. First, a traumatic stressful event or events might trigger behavioral or biological processes that contribute to the *onset* of a disease process. After this process is set into action, the disease could then develop over many years. In this case, the trigger could occur at any time in the life course. Testing this assumption would require a technique that allows for reporting and timing of traumatic events over several decades, such as appropriately adapted event interviews or checklists. Second, long-term exposure to a chronic stressful experience might facilitate the development of disease during stressor exposure. Exposure to a chronic stressor might result in permanent or at least long-term psychological, biological, and behavioral responses that alter the progression of the disease. This hypothesis assumes exposure to chronic stressor(s) and hence requires measurement of chronic stressor experiences. Alternative types of measures include event interviews focusing on chronic experiences and scales designed to assess specific chronic stressors such as marital discord, unemployment, and disability. Obviously, measures assessing acute stressful events such as daily events, or acute stress perceptions or stress responses such as perceived stress, and negative affect measures would be inappropriate in testing either of these hypotheses. An alternative example is the role of stressors in triggering heart attacks (myocardial infarctions). The assumption in this case is that sudden and acute stressful experiences trigger coronary events (Cohen, Kaplan, & Manuck, 1994). Here, acute measures of life events, perceived stress, and negative affect would be appropriate, whereas chronic strain measures covering months or years would be inappropriate.

Other models are possible, and these are intended only as examples. Our point is that investigators need to understand enough about the course of a disease outcome to be able to generate hypotheses about the *duration and timing* of stressor, appraisal, or stress-response exposure required to influence disease onset or progression. With this information, they can choose a stress measure with a temporal span that matches that of the disease process. Different hypotheses may imply very different types of measures as well as different research designs.

Posing a Specific Question About the Relations Between Stress and Disease

The most important issue in designing a research study is to state clearly the question an investigator wants to answer. The more specific the question, the more guidance it provides for choice of instruments and study design. Global questions—for example, What role do stress-related physiological responses play in overall health?—are important. However, a specific question—for example, Does increased heart rate responsivity mediate the relationship between environmental stressors and the development of cardiovascular disease?—is required for the development of a research protocol and selection of appropriate instruments.

The model in Figure 1.1 suggests a variety of questions one might ask about the relation between stress and disease. In this section, we give examples of some of these questions and of the types of measures one would use for each.

When studying the relation between *environmental demands* and health, the focus is on characterizing the environment or environmental/social changes that influence the onset or progression of disease. As discussed earlier, the majority of research in this area has focused on stressful life events. Event characteristics can include the type or domain of events, the magnitude of events, the temporal characteristics of events, or the nature of relations between combinations of events. Part II of this volume addresses different approaches to assessing environmental demands including checklist life event measurement (Chapter 2), interview measurement of stressful events (Chapter 3), daily and within-day event measurement (Chapter 4), and chronic stressor measurement (Chapter 5). Life event measures, whether checklist or interview, are generally used to assess the *cumulative* impact of events over a prolonged period (ranging from 3 months to a life-time, although these measures most often span a period of 6 months to 2 years). Sophisticated versions of these measures can also provide additional information about the context in which the events occurred to aid in determining the potential impact of the events. For example, in the case of a natural disaster, some people will lose all of their possessions and perhaps even some of their loved ones, whereas others will suffer only minor property loss or be fortunate enough to avoid any loss whatsoever. One would expect this variability to influence the stress perceptions and responses. The sensitivity of event measurement can be improved to the extent that a measurement technique allows this kind of information to be obtained.

Chronic stressor measures generally assess the impact of *individual* stressors that last for prolonged (but often unspecified) periods—for example, enduring economic, work, or marital problems. Although such measures can focus solely on environmental events, they also include assessments of long-lasting appraisals or affective responses to specific environmental characteristics. Finally, daily and within-day event measures are generally used to assess the cumulative impact of the events that occur during a single day. Most of these events (or hassles) are so small that they would not be considered contributors to perceived stress and stress-responses when the other, more chronic approaches to event measurement are employed. Daily event measures have many of the same characteristics of major event measures. For example, they can be used to examine the importance of specified domains of events and to elicit information about the meaning and impact of individual events. They are generally employed in studies of the role of very acute (simultaneous or 1- or 2-day lag) changes in health. However, if administered over long periods of time, they can also be used to assess persistent difficulties and nagging day-to-day problems that continue for long periods of time.

As discussed earlier, appraisals refer to the psychological interpretation of events as demanding, threatening, or challenging as opposed to benign (Lazarus & Folkman, 1984). The study of stress appraisal focuses on what are often termed *vulnerability factors*—the nature of persons and their contexts that make them more or less vulnerable to stress-induced disease. For example, variables such as social support and feelings of control that are thought to contribute to adaptive capacities and hence to protect persons from the pathogenic effects of stressful events are thought to act by short-circuiting the appraisal of stress. Alternatively, variables like type A behavior pattern are thought to increase vulnerability to stressful event-induced disease by accentuating stressor appraisal (Cohen & Edwards, 1989). One can also address whether the type of (or reason for) appraisal plays a role in disease risk. For example, events can be viewed as stressful because of the excessive demands they represent, potential bodily or psychological harm, or because of a high level of challenge (Lazarus & Folkman, 1984). Like stressor measures, measures of appraisal may also assess cumulative perceptions of stress combining responses to multiple events or may assess responses to specific individual events. Measurement of stressor appraisal is discussed in Chapter 6.

Environmental demands that are appraised as stressful are generally thought to influence disease risk through negative emotional responses. Although this association is commonly accepted, there are many interesting unanswered questions about the role of emotion in the stress process. For example, do different stressors produce different emotions? Do the same stressors produce different emotions in different people? Are certain types of emotional response associated with certain types of biological responses or diseases? Do positive and negative emotional responses have the same or opposite effects on biological responses and disease risk? There are several alternative approaches to categorizing emotions that are represented by different measurement techniques. For example, emotions can be categorized on the basis of valence (positive or negative), level of associated arousal, or emotional states such as anxiety, depression, and anger. Choice of the appropriate scale should be based on how close a specific breakdown approximates the question about emotional response that the investigator is concerned with. For example, a study of the role of stress in depressive disorder may focus on a measure of depressed mood, whereas a study of emotion as a pathway linking events to immunity or physical disease may require an assessment of the whole range of emotions. A discussion of the different approaches to the measurement of emotional response to stress is presented in Chapter 7 of this volume.

Similar questions can be asked about physiological responses to stress. Do different stressors have similar influences on responses of the cardiovascular, endocrine, and immune systems? Do the same stressors produce different physiological responses in different people? Do different kinds of appraisal produce differential physiological responses? Endocrine and cardiovascular measures have traditionally been used as biological indicators of stress. As discussed earlier, a variety of hormones have been found to be altered by stressors, appraisals, and negative affective responses, and these hormones have been implicated in the pathogenesis of a wide range of mental and physical disease outcomes. However, there are tremendous differences in the characteristic response of different hormones and in their potential importance for specific disease outcomes. One important response characteristic that is essential in choosing a hormonal measure is the temporal course of the response. Some hormones respond quickly or return to baseline quickly, whereas others have a delayed response or slow recovery. Chapter 8 discusses issues in choosing hormone measures. Cardiovascular measures also differ in their importance as potential precursors of different cardiovascular disorders and in their sensitivity and course of response. These measures are discussed in Chapter 9. Immune measures have not traditionally been viewed as indicators of stress. In fact, it is only recently that stressors, appraisals, and negative affect have been closely tied to immune response (see reviews by Herbert & Cohen, 1993a; 1993b). We include a chapter on immune response here because of the increased interest in the role of stress associated immune changes in physical and psychiatric disorders (Rabin et al., 1989). Individual immune measures differ substantially in both their associations with other categories of stress measures, and their implications for disease susceptibility. Chapter 10 discusses these differences and suggests appropriate criteria for choosing immune measures.

Up to now, our discussion has focused primarily on the questions that can be asked within each of the traditional perspectives. However, our hope is that the model we have presented will encourage broader views of the stress process. In particular, it suggests the possibility of studies integrating multiple perspectives. At the simplest level, one could test the basic assumptions of the model. For example, do stressful events that result in perceptions of stress increase disease risk through changes in biological responses and health practices? Studies combining perspectives have the potential to answer many of the most important questions about the role of stress in disease.

Methodological and Practical Considerations

We feel that it is inappropriate to discuss measurement as if it were the most important issue in designing studies of the relation between stress and illness. We consider appropriate and unbiased sampling, study designs that maximize the ability to make causal inferences, and appropriate statistical analyses equally important criteria for any study (e.g., Cohen et al., 1986; Kessler, 1987). Our choice, in this volume, to focus on measurement is based on our perception that the literature has failed to deal adequately with issues involved in the appropriate selection of stress measures and that compendia of available instruments are lacking. Our focus on measurement, however, does not imply that we feel that measurement is more important than any of these other considerations. Good measurement does not compensate for weak or poor study design.

Because of limitations of space and our desire to maintain a coherent focus on measurement, we cannot adequately address the issues involved in designing studies of stress and disease. However, there are design issues that influence the choice of a measure and hence need to be raised in this context. The central concern is the importance of the interplay of the stress measure and the outcome. We introduced this problem earlier in the discussion of the need to match the temporal courses of the stress measure and the disease. However, there are other issues as well. A common problem is the use of stress and outcome measures that assess closely related if not identical concepts. This overlap (or confounding) is often a problem in studies of psychiatric symptoms since measures of symptoms and stress appraisal often include similar if not identical items (Dohrenwend & Shrout, 1985). Similarly, stressful life event scales often include serious illness and hospitalization as events. These (and other) items overlap with outcome measures assessing serious physical or psychiatric illness (Rabkin & Struening, 1976). The possibility of this type of bias is maximized in cross-sectional designs, and minimized in prospective designs where a stress measurement is used to predict changes in the outcome from the point

of measurement to a follow-up. It can also be minimized by selecting stress measures that are not confounded with an outcome or by carefully removing items from
a scale that are potentially confounded with an outcome. This last procedure must
be done with great care since dropping items may alter the psychometric characteristics of the scale or bias the domain that the scale assesses. For example,
removing hospitalization decreases the overlap of the life event and disease measures, but also results in the omission of a potentially important and impactful
stressful event.

Another problem is the potential for shared response biases that occur when the
same type of measurement method is used to assess both stress and outcome. Take,
for example, paper-and-pencil self-reports. These types of measures are influenced
by a range of idiosyncratic response biases such as responding positively and
avoiding extreme responses. If this method is used in assessing both stress and
outcome, the relation between these measures will be artificially inflated because it
will reflect the identical bias tendencies picked up by both measures. This problem
can be avoided by using different methods—for example, a paper-and-pencil stress
measure and biological or behavioral outcome.

Realistically, practical limitations and limited resources often prevent the use of
measures of stress that one would choose based solely on the conceptual and
methodological issues discussed earlier. Primary among these considerations is the
amount of *time required from subjects*. Studies employing multiple measures, those
conducted over the phone, and those with samples to whom there is limited access
or among whom there is limited interest, often require stress measures that take a
minimal amount of time. As a consequence, interview measures and detailed self-
report measures may not be appropriate in these cases. *Cost of administration* is
another important consideration. For example, although interview measures of
stressful events provide more sensitive measurement of events and their contexts,
they require hours of effort by trained personnel, including the interview process
itself as well as labor-intensive coding of materials. Choices based on the cost and
time of instrument administration require the researcher to balance these constraints
with the need for measurement accuracy.

Moderators of the Relation Between Stressors
and Disease Risk

Correlations between life events and illness have rarely risen above .30, suggesting
that life events may account for less than 10 percent of the variance in illness. A
number of investigators have proposed that relations between stress and illness vary
with preexisting vulnerability factors (see reviews by Cohen & Edwards, 1989;
Cohen & Wills, 1985; Kessler & McLeod, 1985). That is, differences in social
support systems, skills, attitudes, beliefs, and personality characteristics render
some persons relatively immune to stress-induced illness and others relatively susceptible. Although this book does not directly address the issues of moderation,
interest in moderation does have implications for the choice of stress measurement.
As we noted earlier, moderators are often thought to act through their effects of

stress appraisal. For example, a person may not view potentially threatening events as stressful if they believe that their social network will aid them in coping. However, moderators could act at any stage in the stress process. For example, social networks may enforce norms about both emotional and behavioral responses to events that are appraised as stressful. By specifying the point in the stress process where moderation is thought to occur, one can choose the type of stress measure most likely to demonstrate moderation. Again, the choice of the appropriate assessment tool depends on the questions under consideration.

Purpose and Organization of the Book

The purpose of the proposed volume is to serve as a resource for state-of-the-art assessment of stress in studies of physical and psychiatric illness in humans. Our major goal in producing the volume is to aid researchers in making decisions about the appropriate measures to use in specific studies. In this chapter we provided a broad conceptual overview of stress and its relation to psychiatric and physical disorder. We have addressed the conceptualization of stress and how environmental, psychological, and biological approaches can answer different questions about the stress process. The overall theme of this chapter and of the volume is that the measures one selects should reflect the kinds of questions being asked and research designs employed. In Parts II through IV of the volume, we address three broadly defined types of stress measurement: environmental demands, psychological stress, and biological stress. We do not attempt to cover all the measures that might broadly be construed as falling in these categories. Instead we choose those that we believe will be of interest to the broadest possible audience. For example, we do not discuss measures of physical environment such as noise and air pollution, behavior changes such as deficits in task performance, or maladaptive coping responses such as smoking and drinking, or alternative biological stress responses such as skin conductance, eye movement and pupillary diameter, or muscle tension. The exclusion of these measures does not imply that they are not as important as or more important than those we discuss. It merely reflects our decision to address a few of the most central categories of measures in depth rather than try to cover all possible measures superficially.

Each chapter provides a conceptual underpinning of the approach it addresses, discusses the important measures within the approach, the kinds of studies suited to each, and the various costs and benefits of using each alternative measure. The purpose of these chapters is to aid researchers in making decisions about the appropriate measures to use in specific studies. Questions addressed in each chapter include: What kinds of questions about the relation between stress and illness can be answered using this kind of measure? Under what conditions would an investigator use one version of a measure versus another? What are the kinds of study designs in which this particular measure works well? What are the logistical issues that must be considered in using this measure? What is known about the psychometrics of these kinds of measures (and, when appropriate, specific measures)? Are these measures population sensitive, and are appropriate measures available for different populations? What are the key references for finding the measures, the psychometrics, and instructions for use? Our intent is to provide valuable information for audiences with a wide range of expertise; to aid persons without extensive experience, but at the same time to provide sufficient information for experts to select state-of-the-art measurement instruments.

References

Ader, R. (1980). Psychosomatic and psychoimmunological research. Presidential Address. *Psychosomatic Medicine, 42,* 307–321.

American Psychiatric Association. (1987). *Diagnostic and statistical manual of mental disorders* (rev. 3rd ed.). Washington, DC: Author.

Anisman, H., & Zacharko, R. M. (1992). Depression as a consequence of inadequate neurochemical adaptation in response to stressors. *British Journal of Psychiatry, 160,* 36–43.

Appley, M. H., & Trumbull, R. (Eds.). (1967). *Psychological stress: Issues in research.* New York: Appleton-Century-Crofts.

Baum, A., Cohen, L., & Hall, M. (1993). Control and intrusive memories as possible determinants of chronic stress. *Psychosomatic Medicine, 55,* 274–286.

Baum, A., Grunberg, N. E., & Singer, J. E. (1982). The use of psychological and neuroendocrinological measurements in the study of stress. *Health Psychology, 1,* 217–236.

Baum, A., Singer, J. E., & Baum, C. S. (1981). Stress and the environment. *Journal of Social Issues, 37,* 4–35.

Beach, S.R.H., Sandeen, E. E., & O'Leary, K. D. (Eds.). (1990). *Depression in marriage.* New York: Guilford Press.

Bolger, N., DeLongis, A., Kessler, R. C., & Schilling, E. (1989). The effects of daily stress on negative mood. *Journal of Personality and Social Psychology, 57,* 808–818.

Brown, G. W., & Harris, T. O. (1978). *Social origins of depression: A study of psychiatric disorders in women.* London: Tavistock.

Brown, G. W., & Harris, T. O. (1989). *Life events and illness.* New York: Guilford Press.

Cannon, W. B. (1932). *The wisdom of the body.* New York: W. W. Norton.

Checkley, S. (1992). Neuroendocrine mechanisms and the precipitation of depression by life events. *British Journal of Psychiatry, 160,* 7–17.

Cohen, S., & Edwards, J. R. (1989). Personality characteristics as moderators of the relationship between stress and disorder. In R.W.J. Neufeld (Ed.), *Advances in the investigation of psychological stress* (pp. 235–283). New York: Wiley.

Cohen, S., Evans, G. W., Krantz, D. S., & Stokols, D. (1986). *Behavior, health and environmental stress.* New York: Plenum Press.

Cohen, S., Kaplan, J. R., & Manuck, S. B. (1994). Social support and coronary heart disease: Underlying psychologic and biologic mechanisms. In S. A. Shumaker & S. M. Czajkowski (Eds.), *Social support and cardiovascular disease* (pp. 195–221). New York: Plenum Press.

Cohen, S., & Matthews, K. A. (1987). Social support, type A behavior and coronary artery disease. *Psychosomatic Medicine, 49,* 325–330.

Cohen, S., Tyrrell, D.A.J., & Smith, A. P. (1993). Negative live events, perceived stress, negative affect, and susceptibility to the common cold. *Journal of Personality and Social Psychology, 64,* 131–140.

Cohen, S., & Williamson, G. M. (1988). Perceived stress in a probability sample of the

United States. In S. Spacapan & S. Oscamp (Eds.), *The social psychology of health* (pp. 31–67). Newbury Park, Cal.: Sage Publications.

Cohen, S., & Williamson, G. M. (1991). Stress and infectious disease in humans. *Psychological Bulletin, 109,* 5–24.

Cohen, S., & Wills, T. A. (1985). Stress, social support and the buffering hypothesis. *Psychological Bulletin, 98,* 310–357.

Conway, T. L., Vickers, R. R., Ward, H. W., & Rahe, R. H. (1981). Occupational stress and variation in cigarette, coffee, and alcohol consumption. *Journal of Health & Social Behavior, 22,* 155–165.

Dohrenwend, B. S., Askenasy, A. R., Krasnoff, L., & Dohrenwend, B. P. (1978). Exemplification of a method for scaling life events: The PERI Life Events Scale. *Journal of Health and Social Behavior, 19,* 205–229.

Dohrenwend, B. P., Raphael, K. G., Schwartz, S., Stueve, A., Skodol, A. (1993). The structural event probe and narrative rating method (SEPRATE) for measuring stressful life events. In L. Goldberger & S. Bresnitz (Eds.), *Handbook of stress: Theoretical and clinical aspects* (2nd ed., pp. 174–199). New York: The Free Press.

Dohrenwend, B. P., & Shrout, P. (1985). "Hassles" in the conceptualization and measurement of life stress variables. *American Psychologist, 40,* 780–785.

Dunbar, H. F. (1954). *Emotions and bodily changes; a survey of literature on psychosomatic interrelationships, 1910–1953* (4th ed.). New York: Columbia University Press.

Eckenrode, J., & Gore, S. (1990). *Stress between work and family.* New York: Plenum Press.

Elliott, G. R., & Eisdorfer, C. (Eds.). (1982). *Stress and human health: Analysis and implications of research; a study by the Institute of Medicine, National Academy of Sciences.* New York: Springer.

Folkins, C. H. (1970). Temporal factors and the cognitive mediators of stress reaction. *Journal of Personality & Social Psychology, 14,* 173–184.

Hawkins, N. G., Davies, R., & Holmes, T. H. (1957). Evidence of psychosocial factors in the development of pulmonary tuberculosis. *American Review of Tuberculosis and Pulmonary Diseases, 75,* 768–780.

Herbert, T. B., & Cohen, S. (1993a). Depression and immunity: A meta-analytic review. *Psychological Bulletin, 113,* 472–486.

Herbert, T. B., & Cohen, S. (1993b). Stress and immunity in humans: A meta-analytic review. *Psychosomatic Medicine, 55,* 364–379.

Herbert, T. B., Cohen, S., Marsland, A. L., Bachen, E. A., Rabin, B. S., Muldoon, M. F., & Manuck, S. B. (1994). Cardiovascular reactivity and the course of immune response to an acute psychological stressor. *Psychosomatic Medicine.*

Herd, J. A. (1986). Neuroendocrine mechanisms in coronary heart disease. In K. A. Matthews et al. (Eds.), *Handbook of stress, reactivity and cardiovascular disease* (pp. 49–70). New York: Wiley.

Herd, J. A. (1978). Physiological correlates of coronary-prone behavior. In T. M. Dembroski, S. M. Weiss, J. L. Shields, S. G. Haynes, & M. Feinleib (Eds.), *Coronary-prone behavior* (pp. 129–136). New York: Springer-Verlag.

Hinkle, L. E., Jr., & Wolff, H. G. (1958). Ecologic investigations of the relationship between illness, life experiences and the social environment. *Annals of Internal Medicine, 49,* 1373–1388.

Holmes, T. H., & Masuda, M. (1974). Life changes and illness susceptibility. In B. S. Dohrenwend & B. P. Dohrenwend (Eds.), *Stressful life events: Their nature and effects* (pp. 45–72). New York: Wiley.

Holmes, T. H., & Rahe, R. H. (1967). The social readjustment rating scale. *Journal of Psychosomatic Research, 11,* 213–218.

Kessler, R. C. (1987). The interplay of research design strategies and data analysis proce-
    dures in evaluation of the effects of stress on health. In S. V. Kasl & C. L. Cooper
    (Eds.), *Stress and health: Issues in research methodology* (pp. 113–140). New York:
    Wiley.
Kessler, R. C., & Magee, W. (1993). Childhood adversity and adult depression. *Psychologi-
    cal Medicine, 23,* 679–690.
Kessler, R. C., & McLeod, J. D. (1985). Social support and mental health in community
    samples. In S. Cohen & S. L. Syme (Eds.), *Social support and health* (pp. 219–240).
    New York: Academic Press.
Kessler, R. C., Price, R., & Wortman, C. (1985). Social factors in psychopathology: Stress,
    social support, and coping processes. *Annual Review of Psychology, 36,* 531–572.
Krantz, D. S., Glass, D. C., Contrada, R., & Miller, N. E. (1981). *Behavior and health.*
    (National Science Foundation's second five-year outlook on science and technology).
    Washington, DC: U.S. Government Printing Office.
Laudenslager, M. L. (1988). The psychobiology of loss: Lessons from humans and nonhu-
    man primates. *Journal of Social Issues, 44,* 19–36.
Lazarus, R. S. (1966). *Psychological stress and the coping process.* New York: McGraw-
    Hill.
Lazarus, R. S. (1975). A cognitively oriented psychologist looks at biofeedback. *American
    Psychologist, 30,* 553–561.
Lazarus, R. S. (1977). Cognitive and coping processes in emotion. In A. Monat & R. S.
    Lazarus (Eds.), *Stress and coping* (pp. 145–158). New York: Columbia University
    Press.
Lazarus, R. S. (1981). The stress and coping paradigm. In C. Eisdorfer, D. Cohen,
    A. Kleinman, & P. Maxim (Eds.), *Models for clinical psychopathology* (pp. 177–
    214). New York: Spectrum.
Lazarus, R. S., & Folkman, S. (1984). *Stress, appraisal, and coping.* New York: Springer.
Levi, L. (Ed.). (1972). *Stress and distress in response to psychosocial stimuli.* New York:
    Pergamon Press.
Lief, A. (Ed.). (1948). *The commonsense psychiatry of Dr. Adolf Meyer.* New York:
    McGraw-Hill.
Manuck, S., Olsson, G., Hjemdahl, P., & Rehnqvist, N. (1992). Does cardiovascular
    reactivity to mental stress have prognostic value in postinfarction patients? A pilot
    study. *Psychosomatic Medicine, 54,* 102–108.
Mason, J. W. (1975). A historical view of.the stress field, Part 2. *Journal of Human Stress, I,*
    22–36.
McCubbin, J. A., Richardson, J., Obrist, P. A., Kizer, J. S., & Langer, A. W. (1980).
    *Catecholaminergic and hemodynamic responses to behavioral stress in young adult
    males.* Paper presented at the Twentieth Annual Meeting of the Society for Psycho-
    physiological Research.
McGonagle, K. A., & Kessler, R. C. (1990). Chronic stress, acute stress, and depressive
    symptoms. *American Journal of Community Psychology, 18,* 681–705.
McGrath, J. E. (Ed.). (1970). *Social and psychological factors in stress.* New York: Holt,
    Rinehart and Winston.
Mechanic, D. (1972). Social psychologic factors affecting the presentation of bodily com-
    plaints. *New England Journal of Medicine, 286,* 1132–1139.
Mechanic, D. (1977). Some modes of adaptation: Defense. In A. Monat & R. S. Lazarus
    (Eds.), *Stress and coping* (pp. 244–257). New York: Columbia University Press.
Meyer, A. (1951). The life chart and the obligation of specifying positive data in psycho-
    pathological diagnosis. In E. E. Winters (Ed.), *The collected papers of Adolf Meyer,*

Vol. 3: Medical teaching (pp. 52–56). Baltimore: The Johns Hopkins University Press.

Murrell, S., Norris, F., & Hutchins, G. (1984). Distribution and desirability of life events in older adults: Population and policy implications. *Journal of Community Psychology, 12,* 301–311.

Neilson, E., Brown, G. W., & Marmot, M. (1989). Myocardial infarction. In G. W. Brown & T. O. Harris (Eds.), *Life events and illness* (pp. 313–342). New York: Guilford Press.

Newcomb, M., Huba, G., & Bentler, P. (1981). A multidimensional assessment of stressful life events among adolescents: Derivations and correlates. *Journal of Health and Social Behavior, 22,* 400–414.

Paykel, E. S. (1974). Life stress and psychiatric disorder: Applications of the clinical approach. In B. S. Dohrenwend & B. P. Dohrenwend (Eds.), *Stressful life events: Their nature and effects* (pp. 135–149). New York: John Wiley.

Pennebaker, J. W. (1982). *The psychology of physical symptoms.* New York: Springer-Verlag.

Raab, W. (1971). Cardiotoxic biochemical effects of emotional–environmental stressors—fundamentals of psychocardiology. In L. Levi (Ed.), *Society, stress, and disease* (Vol. 1, pp. 331–337). London: Oxford University Press.

Rabin, B. S., Cohen, S., Ganguli, R., Lysle, D. T., & Cunnick, J. E. (1989). Bidirectional interaction between the central nervous system and the immune system. *CRC Critical Reviews in Immunology, 9,* 279–312.

Rabkin, J. G., & Struening, E. L. (1976). Life events, stress and illness. *Science, 194,* 1013–1020.

Rahe, R. H., & Lind, E. (1971). Psychosocial factors and sudden cardiac death: A pilot study. *Journal of Psychosomatic Research, 15,* 19–24.

Sachar, E. J. (1975). Twenty-four hour cortisol secretion patterns in depressed and manic patients. *Progress in Brain Research, 42,* 81.

Sandler, I., & Ramsay, T. (1980). Dimensional analysis of children's stressful events. *American Journal of Community Psychology, 8,* 285–302.

Sarason, I. G., Johnson, J. H., & Siegel, J. M. (1978). Assessing the impact of life changes: Development of the life experiences survey. *Journal of Consulting and Clinical Psychology, 46,* 932–946.

Schachter, S., & Singer, J. E. (1962). Cognitive, social and physiological determinants of emotional state. *Psychological Review, 69,* 379–399.

Schulz, R., Visintainer, P., & Williamson, G. M. (1990). Psychiatric and physical morbidity effects of caregiving. *Journal of Gerontology: Psychological Sciences, 45,* P181–P191.

Selye, H. (1956). *The stress of life.* New York: McGraw-Hill.

Selye, H. (1974). *Stress without distress.* Philadelphia: J. B. Lippincott.

Selye, H. (1980). Stress, aging, and retirement. *Journal of Mind and Behavior, 1,* 93–110.

Shrout, P. E. (1981). Scaling of stressful life events. In B. S. Dohrenwend & B. P. Dohrenwend (Eds.), *Stressful life events and their contexts* (pp. 29–47). New Brunswick, NJ: Neale Watson Academic Pub.

Stokes, P. E. (1987). The neuroendocrine measurement of depression. In A. J. Marsella, R.M.A. Hirschfeld, & M. M. Katz (Eds.), *The measurement of depression* (pp. 153–195). New York: Guilford Press.

Stone, A. A., Reed, B. R., & Neale, J. M. (1987). Changes in daily event frequency precede episodes of physical symptoms. *Journal of Human Stress, 13,* 70–74.

Thoits, P. A. (1983). Dimensions of life events that influence psychological distress: An evaluation and synthesis of the literature. In H. B. Kaplan (Ed.), *Psychological stress: Trends in theory and research* (pp. 33–103). New York: Academic Press.

U.S. Congress, Office of Technology Assessment (September 1992). *The biology of mental disorders (OTA-BA-538)*. Washington, DC: U.S. Government Printing Office.

Wheaton, B. (1990). Life transitions, role histories, and mental health. *American Sociological Review, 55,* 209–223.

Wolff, H. G., Wolf, S. G., Jr., & Hare, C. C. (Eds.). (1950). *Life stress and bodily disease*. Baltimore: Williams & Wilkins.

PART II

The Environmental Perspective

The environmental perspective focuses on a stimulus-based definition of stress. The goal is to discover the objective environmental conditions that promote stress and lead to disease. There is considerable disagreement regarding the characteristics that define objective conditions (stressors) that place people at risk for disease, and different measurement methods reflect various positions as to the importance of some characteristics versus others. Characteristics receiving the most attention and reflected in common measurement techniques include the amount of change created by the stressor, the extent to which the stressor leads to loss or threat of loss, and the extent to which the stressor and its resolution are under the control of the respondent. Traditional measurement of environmental conditions has also focused on events with different magnitudes of importance and different temporal characteristics. For example, stressful life events represent major changes that range from short-term to enduring; daily events are more minor day-to-day conditions; and chronic events tend to be important and enduring events.

Part II addresses the measurement of environmental stress. Chapter 2 reviews the voluminous literature on checklist measures of life events, whereas Chapter 3 reviews the more recent and circumscribed literature on interview measures of life events. The decision as to which type of life event measure to use is critical, as the two types generally represent different conceptions of what constitutes a stressful experience and perform quite differently. Checklist measures are much easier to administer but are generally less predictive of illness outcomes. Interview measures require the use of more highly trained interviewers, long personal interviews, and elaborate coding schemes to distinguish subjective components from more objective contextual information. This added effort pays off in more powerful prediction of illness outcomes, but it is not clear how large this advantage is in particular cases.

Chapter 4 discusses measures of daily and within-day events. This is a comparatively new area of investigation for stress researchers, but one that is gaining increasing attention as evidence accumulates that these microstressors are important determinants of both minor acute-onset illnesses and acute flare-ups of more serious chronic conditions. Chapter 5, finally, addresses the measurement of chronic stressors. Some types of chronic stressors, such as frequent arguments with spouse, have long been part of life event checklists. However, recent research has shown that these measures need to be considered separately, as they are more important than life events in explaining the onset and course of chronic health problems. The separate measurement of chronic stressors can also help interpret the effects of life events on more acute illness outcomes because these effects are often mediated by the exacerbation of chronic stressors or potentiated by the contexts created by chronic stressors. Unfortunately, there are substantial methodological problems associated with obtaining objective measures of chronic stressors. Chapter 5 reviews these problems, summarizes domain-specific measures that have overcome these problems, and discusses prospects for future developments of more satisfactory broad-based measures of chronic stressors.

2

Checklist Measurement of Stressful Life Events

R. Jay Turner and Blair Wheaton

As noted in the preceding chapter, epidemiological research on the significance of social stress for health and well-being can be traced to the 1930s. However, the onset of the continuing preoccupation among epidemiologists with the role and significance of life stress followed Holmes and Rahe's (1967) publication of the Social Readjustment Rating Scale (SRRS). This instrument elaborated the Schedule of Recent Experiences (Hawkins, Davies, & Holmes, 1957) by employing a panel of judges to rate the amount of adjustment they thought each event would typically require. In retrospect, the most crucial effect of this innovation may have been its contribution to the face validity of the checklist approach. The availability of normatively generated differential weights overcame the counterintuitive necessity of treating events that seemed of very different severity as though they were equally relevant to health and well-being. Whatever the nature of the advance embodied in the SRRS, its publication was followed by a huge outpouring of research, and it became the best known and the most widely used approach to measuring life events.

The conceptualization that initially informed the checklist approach in general, and the procedure for weighting each experienced event in particular, posited that one's level of experienced stress was embodied in the cumulative amount of change or readjustment brought about by events occurring in one's life (Holmes & Rahe, 1967). The idea that the amount of change is the event property that is responsible for the stressful impact of life events was based largely on Selye's (1956) contention that stress is comprised of nonspecific biological change elicited in response to environmental events and that even relatively small changes occurring in close succession can influence susceptibility and thus disease.

The contention that it is change per se, whether positive or negative in nature, that matters, was also forcefully argued by Barbara Dohrenwend (1973). She, along with her colleagues, saw stressful life events as "objective occurrences of sufficient magnitude to bring about changes in the usual activities of most individuals who experience them" (Dohrenwend, Krasnoff, Askenasy, & Dohrenwend, 1982). This conceptualization continues to inform the work of this distinguished team of stress researchers and is clearly reflected in their recently published measure of eventful stress, referred to as the Structured Event Probe and Narrative Rating, or SEP-

RATE, method (Dohrenwend, Raphael, Schwartz, Stueve, & Skodol, 1993). Notably, this measure considers change as only one of a number of stimulus and contextual characteristics of the event that must be considered, reflecting the elaboration of thinking in life event research over the last 25 years.

The major alternative conceptualizations emphasize not change per se, but a range of issues, including the quality of the experience in terms of undesirability (Paykel, Prusoff, & Uhlenhuth, 1971; Paykel, 1979), contextual threat (Brown, 1981, 1989), and personal control over occurrence of the event (Dohrenwend et al., 1993). As described in Chapter 3, contextual threat is estimated by means of detailed analyses based on interview probes and ratings of independent judges rather than checklist procedures. In contrast, life event checklists (either in questionnaire or interview format) remain the dominant method employed by those who view undesirability as the most relevant issue and who believe that stress cumulation has a generic demand component quite apart from content differences among events. On the basis of a substantial body of research, the majority of life event researchers have come to focus upon undesirable change assessed with lists containing putatively negative events.

Over the 25 years since the publication of the SRRS, there has been an active debate over its adequacy and over alternative approaches to measuring life events and to evaluating their stress-engendering properties. Nearly a decade ago Sandler and Guenther (1985) were able to list 15 critical reviews of stressful life event methodology, and similar commentary has continued to appear (e.g., McQuaid et al., 1992; Miller, 1989; Moos & Swindle, 1990; Newcomb, Huba, & Bentler, 1986; Raphael, Cloitre, & Dohrenwend, 1991). However, the checklist method has evolved in response to criticism, and a substantial array of different life event scales have been developed. Zimmerman (1983) provides a listing of 16 published and two unpublished life event inventories, and several more have appeared since (e.g., Barnett, Hanna, & Parker, 1983; Dohrenwend et al., 1993; Swearingen & Cohen, 1985). It is clear, in addition, that many other "hybrid" mixtures of existing inventories are used regularly in research. Zimmerman (1983) is among those who believe that the more recently developed inventories have overcome many of the central criticisms of the checklist method, a view that has been contested by others (e.g., Brown, 1981; Monroe, 1982a, 1982b; Raphael et al., 1991).

Despite several clearly justified criticisms and associated questions about the reliability and validity of resulting scores, the various published and unpublished life event inventories collectively represent the traditional, and still dominant, research procedure for estimating variations in stress exposure. A substantial preponderance of the now massive literature assessing the linkage between social stress and health is based upon checklist measures of life stress. It is this literature that has compellingly indicated a reliable association between life stress and the occurrence of both psychological distress and psychiatric disorder (Kessler, Price, & Wortman, 1985; Myers, Lindenthal, & Pepper, 1975; Paykel, 1978; Tausig, 1982; Thoits, 1983). Similarly, on the basis of a large number of studies in which checklists have been employed, most authorities (e.g., Barrett, Rose, & Klerman, 1979; Dohrenwend & Dohrenwend, 1974; Jenkins, 1976; Minter & Kimble, 1978) have deemed the weight of evidence to indicate that people exposed to stressful life events are at greater risk for physical health problems. For example, Jemmot and Locke (1984) report that individuals exposed to high levels of life stress (as measured by life events) experience "greater degeneration of overall health, more diseases of the upper respiratory tract, more allergies, a greater incidence of hypertension and a greater risk of sudden cardiac death and coronary diseases than do people who have been exposed to a low degree of life stress" (Jemmot & Locke, 1984:79). The significance of life stress has also been reported in relation to tuberculosis, diabetes, arthritis, cancer (Holmes & Masuda, 1974), accidents (Selzer & Vinokur, 1974), athletic injuries (Bramwell, Wagner, Masuda, & Holmes, 1975), and poor academic performance (Lloyd, Alexander, Rice, & Greenfield, 1980).

One conclusion that seems to follow from the consistency with which relationships are observed between event accumulations and distress and disorder is that whatever the shortcomings of the method, event inventories yield estimates of stress exposure that are meaningful. There are essentially two challenges to this conclusion. One is that observed associations are an artifact of operational confounding arising from the fact that many event checklists, perhaps the majority historically, have included items that are themselves symptoms, or indicators, of physical or mental illness or distress, or logical consequences of these problems. The issue is considered below in relation to "event selection." Here we simply note that although some research has suggested that measurement confounding or contamination may account for the persistent association between life event measures and both physical (Schroeder & Costa, 1984) and psychological health (Thoits, 1981), many more studies have demonstrated that the relationships at issue persist, and are little attenuated, when confounded items are removed from consideration (Billings & Moos, 1982; Grant, Sweetwood, Yager, & Gerst, 1981; McFarlane, Norman, Streiner, Roy, & Scott, 1980; Mueller, Edwards, & Yarvis, 1977a; Sandler & Block, 1979; Tausig, 1982).

The second challenge has to do with the possibility of what could be termed "causal confounding," in fact, the possibility of causal reciprocity with outcome variables. Concern that illness or distress also produces either a real or artifactually higher number of reported events has been a continuing theme of the life events literature. This issue needs to be clarified before we proceed. First, the possibility that illness also causes stress exposure does not question the causal status of stress with respect to illness; it simply complicates the proper estimation of this causal impact. The same could be said for other possible variables that act as background confounders related to both stress exposure and later illness: to state the obvious, their impact must be controlled either statistically, or by design. Second, evidence from longitudinal studies consistently suggests that if one controls for prior or baseline distress or mental health, life events have a net impact on change in distress or mental health over time (Aneshensel & Frerichs, 1982; Eaton, 1978; Myers, Lindenthal, & Pepper, 1974; Turner & Noh, 1988; Turner & Roxburgh, in press). Thus, at least with respect to mental health, two essential criteria of causality are supported in the literature: (1) statistical association and (2) temporal or logical precedence. The most important remaining issue is the possibility of spuriousness in the effect of life events on health outcomes. It is much more difficult to establish this criterion, but it is fair to say that the literature is replete with examples of attempts to control for confounding factors in the events–illness relationship and yet precious few examples that actually refute this relationship. Thus, although less than definitive, we believe the evidence is suggestive of the plausibility and even the likelihood that the experience of stressful events is causally implicated in observed variations in health risk.

In this chapter we provide information and make recommendations relevant to fashioning an appropriate life event inventory. The major issues examined include:

1. The definition of events and the distinction between events and other possible sources of social stress.
2. Event selection and the relevance of, and effects on findings of, content differences and number of events comprising the inventory. The significance of the content issue is also considered in relation to age and cultural variations in study populations.
3. Use of change versus presumed undesirability as a basis for defining events to be included on inventories.
4. Operational confounding in event measures.
5. The importance of comprehensiveness of event lists.
6. The effects of variation in role occupancy on event score totals.
7. Time frames for studying eventful stressors.
8. The advisability of, and alternative approaches to, event weighting.
9. Reliability concerns.

The final section presents a summary of our recommendations regarding event checklists and evaluates their adequacy for estimating variations in exposure to social stress. We also comment on some promising innovations that have been applied within event inventories, and argue that a more encompassing checklist procedure can provide dramatically better estimates of stress exposure than event inventories have so far provided. Our discussion assumes throughout that "checklists" refer in fact to both interview-derived and self-administered inventories (since the lists used are often the same), and that inventories form the core starting point of contextually scored measures.

Defining and Selecting Life Events

Since negative change, or rather, the undesirable demand or threat accompanying change, is generally assumed to represent a basis for the experience of stress, we suggest that the universe of *eventful* stress is uniquely characterized by discrete and, in principle, observable environmental and social changes. These changes are potentially threatening because they precipitate the need for adjustment in identity or life routines and are sufficiently important in their potential impact that they cannot be interpreted or treated as relatively routine exigencies of daily life. This definition specifically excludes minor events that are more precisely considered hassles or irritants, insidiously developing life conditions that could be considered stressful, and more enduring or chronic problems.

Our approach recognizes the worth and the necessity of separating sources of stress as distinct factors in the stress process. A basic question is, when does an event stop being an event? The essential components of our definition of event stress are the facts of (1) discreteness, (2) observability in principle, and (3) a self-limiting time course (for the *event*, not necessarily the consequences). Thus, when considering the effects of job loss, we consider job loss the event and any resulting financial strain a separate, chronic stressor, in part caused by the job loss. Note that financial strain can have other causes, and thus it cannot be considered "part" of the event of job loss. In the case of divorce, we would treat any ensuing role overload due to single parenthood a separate stressor, not part of the "event." To fold such consequences into the event would be a mistake for two reasons: (1) It would disallow the separate predictive and explanatory role of other forms of stress; and (2) whether chronic stress follows from the event is an empirically contingent matter and not a necessary part of the process of the event. Some job losses are not followed by financial strain because the spouse goes to work; some divorces do not lead to role overload because a new relationship starts, and so on.

It is clear that the occurrence of some "events" can signal the onset of enduring difficulties that may eventually have independent stress-evoking potential. The question of the health significance of an event, relative to other sources or types of stress, has yet to be convincingly answered; however, it is plausible that some *portion* of the well-established relationship between life events and health derives from associated difficulties or strains consequent to the occurrence of the event.

The above definition, like that previously quoted from Dohrenwend et al. (1982), emphasizes the goal of assessing differences in exposure to social and environmental stress rather than differences in the experience of events based on their personally attributed meanings. Since research subjects self-report events occurring in their lives, there is inevitably a subjective element. However, many stress researchers view the central properties of events as relatively objective in nature. These properties include whether or not the event occurred, to whom the event occurred, and when the event occurred.

In our opinion, there are heuristic advantages to conceptualizing life events as objective or quasi-objective phenomena because failure to do so tends to confound differences in exposure to events of stress-evoking potential with differences in vulnerability to such events. It is not suggested that the individuals' perception and appraisal of experienced events are unimportant. Indeed, we see the personal experience of stress as a joint function of the nature and severity of the event and of the individual's capacity (perceived and real) to handle it in both emotional and practical terms. This latter element, which reflects vulnerability differences that condition the rate at which exposure to stressors is translated into distress or illness, has usually been indexed in terms of such factors as prior experience, possession of relevant coping skills, and availability of personal and social coping resources. Cleary's (1981) position on this issue is appropriate: "An attempt should be made to keep the assessment of . . . events separate from the assessment of various mediating factors and other processes contributing to illness" (p. 311). Only when a clear distinction is maintained between the relatively objective estimation of level of exposure to eventful stressors and reactions to eventful stressors can the respective significance of exposure and vulnerability be evaluated and the potential of intervention efforts in these two domains be estimated. It is important to note that this distinction is jeopardized by efforts to weight inventory items differentially on the basis of respondent reports of the severity or stressfulness of experienced events.

It is evident that the assessment of life events by checklist is hardly a standardized procedure. In terms of item content, a substantial number of different lists have been generated and employed in research but none stands out as definitively superior to the others. Moreover, as several reviewers seem to imply (Cleary, 1980; 1981; Monroe, 1982a; Zimmerman, 1983), a specific definitive list of appropriate events that can be applied across a range of study populations cannot be specified and should not be recommended. Obviously, lists that attempt general comprehensiveness will include items that are not relevant to particular target populations. Examples include items like "retirement," "started menopause," and "spouse died," which are quite irrelevant within an adolescent or young adult population, and items like "failed a grade in school" and "started work for the first time" are of unlikely significance for the elderly. That the inclusion of events that are irrelevant to many respondents is, in most instances, inevitable is demonstrated by the fact that even within a relatively homogeneous population, individuals vary substantially with respect to their occupancy of social roles, and the relevance of many events is tied to role occupancy. It seems clear that the rational requirement that "events listed in a life events inventory must be ones for which the individuals in a study population would be at risk in the time span over which the study respondents would be required to report" (Hurst, 1979:18), can seldom be achieved in general, and never be achieved within general population studies. Given this circumstance, it seems crucial to ensure that the event list incorporates a reasonable and balanced representation of events that are of potential relevance to respondents occupying differing constellations of role sets.

Beyond this requirement, appropriate item content of a selected or devised inventory depends upon the nature of the population and of the outcomes being studied. A primary consideration in this regard, of course, is the age or life stage of the target sample. Recognition of the importance of this dimension has led to the development of checklists tailored specifically for children, adolescents, and elderly populations (see Appendix for examples). In addition, despite early reports indicating that the meaning of events in terms of the adaptation required is consistent across nationalities, cultures, races, and income levels (Holmes & Masuda, 1972; Masuda & Holmes, 1967a, 1967b) evidence that the perceived stressfulness of events differs from culture to culture (Askenasy, Dohrenwend, & Dohrenwend, 1977; Dohrenwend, Krasnoff, Askenasy, & Dohrenwend, 1978; Hough, Fairbank, & Garcia, 1976), suggests that event lists should be tailored to the cultural group and, indeed, to whatever other specific subgroup may be the focus of the investigation (Hurst, 1979). In this connection, the recommendation seems appropriate that events selected from other inventories, based on their apparent relevance, be supplemented by events nominated by representatives of the target group itself, from their own experience and that of others they know (Dohrenwend et al., 1978). As Monroe (1982a:438) has noted, this kind of approach has the disadvantage that it does not allow stress score comparisons across diverse populations; however, "the advantages in terms of instrument sensitivity and relevance suggests its utility over more 'standardized' procedures."

This problem is ameliorated somewhat by one feature of the literature that should be mentioned in reference to event selection. Despite the fact that there are many variants of life event inventories in existence, it is now apparent that content differences do not result in major differences in the magnitude of relationships, at least with respect to well-known mental health outcomes. Whether this is true for physical health outcomes is less clear. In other words, in using life events to predict mental health, concerns about moderate content differences between one inventory and another are probably misplaced, as long as the major events are included. Such events are also likely to be the common core of many inventories.

Change Events Versus Negative Change Events

As noted earlier, there is a continuing debate about whether it is change or undesirability that is relevant to the experience of stress. The resolution of this issue requires a criterion against which to evaluate these alternatives, but the selection of the best criterion in itself merits prolonged debate. However, since the purpose of this volume is to consider measurement strategies for assessing the significance of life stress for physical and mental health, we follow the lead of researchers with similar interests (e.g., Mueller et al., 1977b; Myers et al., 1974; Ross & Mirowsky, 1979; Vinokur & Selzer, 1975) in accepting power to predict the dependent variable as the relevant criterion. This standard, of course, was suggested in Selye's (1976:78) definition of a stressor as "that which produces stress" and has been explicitly or implicitly adopted within many conceptualizations of the stress process (e.g., Pearlin, Lieberman, Menaghan, & Mullan, 1981; Rabkin & Struening, 1976). Since we have no means of directly assessing the experience of stress (biologically defined) in field settings, reports or other indices of distress or illness must be taken as both reflections and consequences of the experience of stress. Kahn and Quinn (1970:51–52, quoted in Ross & Mirowsky, 1979) providing a succinct commentary on the issue, describe stressors as

> a stimulus situation, the properties of which are response inferred. . . . The problem with this definition of stress is its apparent circularity. . . . The solution to this circularity is not to abandon completely the notion that stress is response-inferred, but to regard it as defined by the response of a whole class of organisms rather than by the responses of a particular individual.

Based on the conclusion that stressfulness is effectively indexed by the distress or illness response of collectivities of individuals, considerable evidence on the change versus undesirability issue has accumulated. From a review of 17 studies that examined the relationships of desirable and undesirable events to measures of psychological distress, Zautra and Reich (1983) observed a consistent pattern of clear positive relationships with negative events in contrast to weak and contradictory findings with respect to positive events. Moreover, in those studies that did find positive events to be associated with increased distress, the relationship could no longer be observed when differences in negative events were taken into consideration. Evidence suggesting the dominant significance of negative compared to positive events as risk factors for distress and illness has been consistent across a range of study populations, dependent variables, and event inventories (e.g., Mueller, Edwards, & Yarvis 1977a, 1977b; Ross & Mirowsky, 1979; Vinokur & Selzer, 1975), and we see no basis for not taking both these findings and their practical implications seriously. However, we cannot with confidence conclude that ostensibly positive life changes play no role in either elevating or conditioning stress experiences, because the mechanisms involved may be more complex than research has so far been able to reveal. Our recommendation from available evidence is that the limited time and space considerations that characterize most questionnaires and interviews require that the content of life event checklists be restricted to events that are presumptively undesirable in nature, and that questions on events normally considered as positive be excluded.

Confounding of Events with Outcomes

The issue of operational confounding was referred to earlier in relation to the suggestion that such confounding might account for the persistent but modest associations observed between event scores and various health outcomes. We indicated that the weight of evidence is reassuring on this matter, revealing that confounded events make only a small contribution and that the relationship quite generally persists when clearly confounded and questionable events are excluded from consideration. However, although it is not a viable competing hypothesis, it is nonetheless an issue of significance when selecting from available checklists or assembling events for one's own list.

The problem of employing events that are themselves indicators or symptoms of illness or adaptational problems when assessing the health–stress relationship has been repeatedly raised in relation to the SRRS as well as other checklists. It has even been suggested that 39 of the 43 events contained in the Holmes and Rahe (1967) scale could be viewed as symptoms or consequences rather than precipitants of illness (Hudgens, 1974). Dohrenwend and Dohrenwend (1974, 1978), having considered the issue in some detail, concluded that events included in the typical inventory are of three types: (1) those that are confounded with the subject's psychological functioning; (2) those that reflect physical illness; and (3) those that are independent of the subject's physical and psychological health status. In their view, the question of the etiologic significance of eventful stress could best be addressed by restricting consideration to the third category of fateful or adventitious events.

Although Zimmerman (1983) has forcefully argued that many of the problems that have led to criticism of the SRRS have been corrected in subsequent instruments, he also notes that a number of more recent scales have failed to exclude some obviously confounded events. Our own view is that regardless of how minimal the impact may be on observed relationships to health status, no event should be included that may directly reflect, *in a measurement sense*, an individual's physical or mental health status. To emphasize "in a measurement sense" is to say that events that are confounded purely because the probability of their occurrence is affected by health status should be—indeed, need to be—included. As noted earlier, causality can occur validly in both directions. The fact that the occurrence of a life event may be partially determined by a prior disorder does not disqualify this event, at a further stage of the event–disease dynamic, from having causal impact on later disorder. Such causal reciprocity, if it exists, should be explicitly investigated, not removed from consideration. In discussing such events, Shrout (1981:35) has noted:

> Since these events themselves presumably add to the stress a person experiences, there exists a feedback loop between life events and health status. Clearly, the inclusion of such events . . . would result in the overestimation of the causal impact of stress on illness, yet the elimination of such events . . . would result in the underestimation of stress and consequently the underestimation of the causal impact of stress.

The fact is, however, that inclusion of such events does not lead to overestimation if causal reciprocity is properly modeled or controlled as part of the analytical process (Kessler, 1983). Thus, inclusion of such events is justified.

At the same time, we also believe that in the area of measurement confounding per se, there is little evidence of confounding in the event lists now commonly used for mental health status, although care must be exercised in choosing events to avoid confounding with physical health status.

Comprehensiveness of Event Lists

It is clear that the Holmes and Rahe (1967) checklist, and most subsequent inventories, contain only a sampling or subset of potentially stressful life events. One problem with this approach is that item sampling theory as it is applied to test construction has not been, and cannot be of much assistance. This means first and foremost that there is no basis for defining a universe of relevant events. As far as we know, no investigator has reported a selection procedure in which every possible event of stress-evoking potential had an equal chance of selection, and few would claim to have identified the full range of events from which a representative sample could be drawn. More importantly, the development of an events checklist is conceptually distinct from test construction because the items are specifically not alternative estimates of a single underlying construct, characteristic, or experience. Since it is not *necessarily* the case that the experience of one event increases the likelihood of another, there should be no expectation that event inventories display internal reliability as estimated by Cronbach's (1951) alpha. On the contrary, high coefficient alpha values and high correlations between individual items, rather than constituting a desirable circumstance, may signal problems with item redundancy, as Pugh and colleagues (1971) have noted, or of both item redundancy and causal dependency among items, as Cleary (1981) has suggested.

On the other hand, there are grounds for anticipating nonzero or even modest intercorrelations, among at least a subset of eventful stressors. This expectation is associated with a perspective recently enunciated by Pearlin (1989), who effectively argued the need for attention to the structural contexts of people's lives. In his view, stressful occurrences and circumstances are among those experiences that are rooted in these contexts, and thus, there is a basis for assuming that variations in exposure to stressors arise substantially out of contemporary and developmental conditions of life. Expressing a similar position, Aneshensel (1992) has called for a conceptual reorientation away from the view of stress as an isolated risk factor and toward its consideration as a link in a chain that begins with social conditions and ends with differences in risk for distress and illness. She draws a distinction between systemic stressors and random stressors. Random stressors are those that are not embedded in social location or social experience over the life course, they occur with roughly equal probability across the various social groups that are known to differ in their risk for mental and physical health problems. Events such as "friendship ended," "miscarriage," or "family member dies" are relatively independent of most social statuses.

Although such stressors are important as predictors of health risk, they are not helpful in explaining established linkages between social conditions and such risk. Systemic stressors are those that are tied to social location and/or social group experience, and thus are directly relevant to understanding what ties health to the social conditions of life. Obviously, events such as losing a job, foreclosure of a loan, going on welfare, being assaulted or robbed, and divorce tend to be linked to one's location in the social system. To the extent that risk for some portion of listed events is associated with ongoing conditions of social life, intercorrelations among such items will be observed. This means that such intercorrelations have a substantive rather than measurement meaning, and should be interpreted as such, perhaps indicating as well internal causal sequences among sets of events.

From a conceptual perspective, the central basis for preferring a comprehensive approach has been clearly described by Shrout (1981:34), who notes that life event lists must be complete because

> if any event that adds to stress is not included, our estimate of stress for persons who experience that event will be inaccurate. From this perspective, the notion of a "sample of life events" is misapplied to research on stress. Since life events are largely independent of one another, we cannot make any inferences about some life events based on information about others for a given person.

In Shrout's opinion, failure to include relevant events may result in substantial error in estimates of stress exposure, and such error may produce a significant underestimation of the relationship between stress exposure and health outcomes.

We agree that there is no theoretical or logical basis for excluding items for which there is evidence that risk for distress or illness is elevated among those exposed, regardless of how rarely the event occurs in the target population. Moreover, we subscribe to the view that a relatively comprehensive approach is advisable for studies that aspire to improve measurement or focus on the intricacies of the stress process. However, for those interested in straightforward estimation of the effect of eventful stress, considered as an aggregate total, our advice is quite different. We suggest a list of 30 to 50 reasonably prevalent events drawn from extant checklists and supplemented as necessary to ensure relevance and appropriateness for the population being studied. This recommendation is based on the apparent absence of any compelling evidence suggesting that longer, more comprehensive lists yield substantially different findings, and considering practical data acquisition requirements associated with time, costs, and respondent burden. In other words, despite the seemingly obvious truth of Shrout's position, evidence seems to suggest an empirical "upper limit" to associations of eventful stress and health outcomes, as if there exists a core of potent items that are most relevant and a similar proportion of "dead weight" items regardless of the length of the inventory.

The tension between the need for comprehensiveness and practicality can be reduced somewhat by regular use of follow-up questioning after a given checklist about other important events that have happened over the target time period. We have found that about half of the responses to this kind of follow-up are codable and recognizable as life events, whereas the other half are clearly symptoms, daily hassles, or chronic stressors. Investigators can choose to supplement their life event totals with these postcoded events and compare results with the predefined list. We find that this practice either allows quite rare events to have an impact or reveals events that are essentially replicated in the inventory but are worded slightly differently. We also suggest that an open-ended question cannot be used in place of specific event lists, because of the increased chance of differential recall bias and increased coding unreliability in open-ended answers. A conservative approach would be to use an open-ended question only as a supplement, and to code only an event that is likely to be mentioned with equal probability by most respondents who have actually experienced the event.

Role Occupancy and Event Exposure

Problems in interpretation of the effects of event totals due to variation in role occupancy are rarely mentioned in the literature (recent exceptions include Raphael et al., 1991, and Aneshensel, 1992). We wish briefly to point out why such problems require consideration. When, for example, respondents do not work, by definition they are not at risk for conflict with the boss, or being demoted or being fired; they are not at risk for divorce or other relationship difficulties if they are unmarried or have no partner, and only those with children can experience difficulties with, and the difficulties of, their children. As a result, respondents in fewer social roles are likely to have lower stressful event scores.

The implications of this confounding are not entirely clear, but one argument is that respondents with more role occupancies (worker, partner, parent, etc.) are in those roles because of generalized social competence, and as such, life event totals will be confounded with social competence. If this is the case, the problem is that this is a positive correlation, but social competence will be negatively correlated with distress/disorder. This combination implies a suppressor (i.e., negative) component in the otherwise positive correlation between events and health problems, thereby reducing the size of this positive correlation. This possibility has rarely been considered. Fortunately, this possibility can be investigated by counting and constructing either role occupancy (as in a count of roles) or role configuration measures (Thoits, 1987), and adjusting for such confounding in the estimation of impacts of events on health outcomes.

In longitudinal studies, if one wants to control for "at risk" status in assessing the effects of events, it is important to remember that role occupancy should be coded from the beginning of the observation period, not at the end. Thus, for example, if the interest is in the effects of widowhood, one must use baseline status and select a group of married respondents.

Event Timing, Interval, and Duration

Although most studies on the health effects of eventful stress have employed a 1-year time frame for the occurrence of relevant events, the maximum lag time considered has varied from a few weeks to several years. Aside from convenience in questioning respondents, the original rationale for the 1-year time frame appears to have been the assumption that the effects of increased stress tend to become manifest in about a year's time (Holmes, 1979; Holmes & Masuda, 1974). However, as Monroe (1982a) has observed, the evidence for this assumption is limited and methodologically questionable (Dohrenwend et al., 1978). Moreover, most of the events obtained on the basis of a 1-year interval will have occurred within a time frame less than 1 year before the outcome of interest. Evidence suggesting the appropriateness of alternative time frames appears split, pointing to intervals on both sides of the 1-year standard. Thus, Brown and Harris (1978) reported that the lag time between the experience of events and the onset of clinically significant depression appeared to be less that 6 months. Consistent with this and other evidence (Murphy & Brown, 1980), Turner and Roxburgh (in press) found, from a three-wave panel analysis, that the effects of eventful stressors were limited to a 6-month interval. In contrast, investigations employing 2, 4 and 5-year time frames (Eaton, 1978; Thurlow, 1971; Wyler, Masuda, and Holmes, 1971) have reported higher correlations than many studies that have considered shorter periods.

If forced to suggest a single time-frame boundary, we see a minimum of a 1-year reference period as appropriate with respect to most events that are typically assessed using the checklist strategy. It is long enough to obtain a reasonable estimate of variations in exposure to recent life events, and short enough to avoid the substantial decline in the ability of respondents to recall events that appears to occur beyond the 1-year time frame. A more general recommendation, however, is that the appropriate time frame will necessarily depend on the nature of the outcome and the inherent causal lag of events for that outcome. In addition, one can investigate the decline in event reporting as a function of months and develop cut-points for a time frame based on the results.

It is important to recognize that the problem of reliability of reporting or remembering, and the conclusion that stressors may be very time limited in their effects, cannot be applied equally to all events that may be experienced. For example, it would rarely be the case that a subject would not remember, or fail to report in response to specific questions, that his or her parent had died, or child had died, or parents had divorced; that he or she had served in combat, or had been physically or sexually assaulted, or had been abused as a child.

There is evidence that a number of traumatic events or circumstances such as these are relevant to mental health and that the duration of their effects is closer to lifetime than to 6 months or 1 year (Rutter, 1989). For example, a substantial body of evidence suggests the possible long-term psychiatric significance of parental loss (e.g., Faravelli, Ambonetti, & Pallanti, 1986; Tennant, 1988), perhaps through the well-established risk factor of maternal lack of care (Birtchnell, 1988; Parker & Hadzi-Pavlovic, 1984). Research also indicates the significance for adult mental health of childhood experiences of parental conflict and divorce (Lauer & Lauer, 1991; McLeod, 1991) and of physical or sexual abuse (Brown & Anderson, 1991; Bryer, Nelson, & Miller, 1987). Kessler and Magee (1993) have recently reported on the long-term relationships between eight forms of childhood adversity and adult depression. They found seven of the eight to be directly associated with early onset of depression. In this study, early trauma is related to adult depression because it has a relatively immediate effect on mental health which either has stability through the life course or is recurrent in terms of episodes. Several rather dramatic and severe events such as the death of a child or the untimely death of one's spouse have, of course, often been incorporated in event lists. However, the rarity of such events within the 6 months or 1 year covered by the typical inventory has inevitably masked their potential health significance.

Thus, there are a range of events that, presumably, can be reliably measured and that can have significant mental health consequences despite their occurrence years or even decades earlier. These twin facts argue strongly for the systematic inclusion of lifetime experience of such events as part of the effort to assess differences in exposure to eventful stressors. Within several studies in our ongoing research program we have incorporated a supplementary checklist of 20 lifetime events or traumas. In addition to ascertaining whether each event had been experienced, subjects are asked their age at first and last occurrence. Preliminary analyses reveal the majority of these major lifetime events to be associated with significant elevation in risk for a lifetime diagnosis of major depressive disorder.

Both the available research literature and our own experience justify the recommendation that a separate list of lifetime events or traumas be added to inventories of recent events. Checklist events for which there is evidence of long-term health effects should be transferred to the lifetime traumas list to avoid a temporal sample that is limited to a single year. Analyses should consider both the direct health effects of lifetime traumas and the extent to which they may condition the impact of more contemporaneous stressors.

Duration of Recent Life Events

As the present volume explicitly acknowledges, it is well recognized that both discrete life events and relatively enduring problems are relevant to distress and illness. As Pearlin and colleagues (Pearlin, 1983; Pearlin et al., 1981) have suggested, the experience of eventful stress may alter the meanings of existing strains, generate new strains, or magnify existing strains. Alternatively, the impact of recent events may be amplified in the presence of chronic stress (Brown & Harris, 1978; Paykel, 1978). The testing of such ideas, however, must confront the problem of conceptually and operationally disentangling eventful stressors from events that signal the onset of an enduring source of stress. Some occurrences of a given event may be discrete or time-limited in nature, whereas other occurrences of the same event may represent the beginning of a long-term difficulty.

One strategy for distinguishing such important differences is that applied by Turner and Avison (1992; Avison & Turner, 1988), who obtained information for each reported event on when it ended as well as when it had occurred. Analyses confirmed their suspicion that life event checklists routinely provide information on differences in exposure to more typically enduring, as well as more discrete, stressors. They found that the majority of all events reported by their large community sample ($n = 1561$) to be discrete in nature, with nearly 70 percent beginning and ending in the same month, and more than 76 percent lasting less than 3 months. However, a substantial proportion were more chronic, with 19.3 percent spanning 10 months or longer. It is important to note that although events that intuitively seem more enduring (e.g., problems with children, problems with one's parents, and financial difficulties) were substantially overrepresented among long-lasting events, the majority of occurrences of each of these were reported to have been time-limited or discrete in nature. Moreover, none of the 31 negative events assembled by these investigators from established lists were found to be uniformly experienced as either discrete or chronic stressors. Thus, it is clear that no reported association between checklist scores and health outcomes reflects the effects of purely discrete events.

Having distinguished discrete from enduring stressors reflected in checklist responses, Avison and Turner (1988) were able to show that both have important independent effects on psychological distress, and that a more effective estimate of an individual's burden of stress at a given point in time is achieved by considering both recently occurring events and the enduring stressors to which the individual is still subjected—a concept we have referred to in our research as the "operant burden" of stress (Turner, Wheaton, & Lloyd, in press). Since it appears that events are salient in terms of their stress-evoking potential for varying periods of time, and that such variations are relevant to health outcomes, effective measurement requires information about the onset and conclusion of the stressful experiences tapped by event checklists. We strongly recommend the inclusion of probes to obtain this important information on all reported events.

As detailed elsewhere in this volume, there is substantial basis for attaching significance to chronic stress (e.g., Liem & Liem, 1978; Ross & Huber, 1985) and to support the contention that it may be of primary significance for mental health (Pearlin & Lieberman, 1979; Wheaton, 1991). The research just reviewed leaves little doubt that enduring stress is a part of what is reflected in associations of checklist scores with health outcomes and that it is possible to distinguish its contribution. However, it is equally clear that an event checklist, regardless of whether probes are included to identify continuing stressors, is too limited and too narrow in content to provide even a remotely adequate estimate of an individual's recent or ongoing exposure to chronic stressors. Wheaton (1991) has recently offered a 51-item inventory that assesses a wide range of role-related and role-unrelated chronic stressors. Elsewhere we have shown that when scores from this inventory are combined with the burden of stress estimated from the events checklist, the variance in depressive symptomatology and in major depressive disorder accounted for by stress exposure is about 2 1/2 times greater than has typically been reported in the literature (Turner et al., in press). Our point is not to recommend one or another approach to measuring chronic stress; that is the task of a separate chapter. It is rather that, because conventional event inventories estimate exposure to eventful stressors and partially estimate exposure to more continuing forms of potentially stressful experience, checklists for measuring social stress are clearly incomplete unless they incorporate a separate sampling of items addressed to the general domain of chronic stress.

Event Weighting

Despite the considerable attention paid to differential weighting of events in the literature, the issue is still not settled, and we believe that some attention to this issue is still necessary. There are many types of weighting, but the two most common are "objective" ratings of the change, or importance, or seriousness, of events by "judges," selected in random samples of defined populations, in purposive samples of individuals who have experienced the event being weighted, or by expert panels, and "subjective" ratings assigned by individuals to their own events and considered at the individual level only. Perhaps the basic difference between the two is the level of aggregation of the weight: Objective weights reflect an average assigned by a group of raters; subjective weights involve self-reporting by individuals as to how serious the event was to them.

Despite repeated and widespread attempts to prove otherwise, the best conclusion from the existing research concerning the effectiveness of differential weighting *using current approaches* is that weighted indices do not generally increase the correlation with outcomes, whether using objective or (surprisingly) subjective weights (Monroe, 1982a; Sandler & Guenther, 1985; Zimmerman, 1983). On the other hand, as nicely pointed out by Zimmerman (1983), the use of unweighted indices puts the life event literature in the absurd position of implying no differences in impact potential across events.

It is not generally realized that different weighting schemes have implicitly contrasting goals. Averaged "objective" weights derived from samples of raters are attempts at capturing differences in the characteristics of *events* that imply differences in impact potential that apply across individuals (Thoits, 1983), whereas subjective weights attempt to capture differences in the impact potential associated with variations in the "meaning" of an event, across individuals. Forced to choose, we prefer the former logic. There is nothing inherently wrong with specifying group-averaged weights, with the understanding that all they can capture, all they *should* capture, is the average differences in impact potential.

Concern about subjective weights most often centers on the issue of confounding of events with outcomes when events are subjectively weighted. First we note that in longitudinal data, it may be possible to take into account this kind of confounding by controlling for time 1 health status, when using time 1 or time 2 events to predict outcomes at time 2. The effect of health status on event scores is partialed out of the effect of events on later health, and thus the bias is reduced.

But our concern with subjective weights applies to quite another issue: the confounding of stress exposure with differential vulnerability to stress, in other words, the confusion of stress with coping. We contend that the weight attached by a respondent to an event will be largely a function of his or her capacity, real or perceived, to resolve that event in emotional or practical terms. The stress process model argues appropriately that such capacity is a function of coping skills and of the availability of social and personal resources. Thus, subjective ratings confound the effects of stress mediators with those of stress exposure and should therefore be discouraged even in prospective studies and studies on physical health status.

A preferable approach is to allow the issue of differences in impact potential to be conceptualized, albeit indirectly, via the specification of coping resources and use of coping strategies. These factors should be considered as separate factors in the explanation of health outcomes, and the ways in which these factors combine with stress events to predict outcomes should be investigated as part of estimating the correct model (e.g., additive burden vs. interactive stress-buffering).

Returning to the issue of objective weights, it may be the case that the whole issue has been laid to rest prematurely. Over 10 years ago, a series of publications provided both advice and clues that at least one approach to weighting—regression-based weighting in which the effects of individual events are weighted in the total index by their effects on the outcome—did provide much larger correlations with outcomes than other approaches (Ross & Mirowsky, 1979; Shrout, 1981; Tausig, 1982). There are basic objections to this approach: that it is atheoretical, and that it leads to nongeneralizable weights. We believe these criticism to be at least potentially misleading in one crucial respect, although correct in other respects.

Shrout (1981) shows that the correlation between weighted and unweighted indices is directly a function of the standard deviation of the weights, relative to the mean of the weights. He argues further that regression-based weighting is an attractive approach, but requires very large samples. This is indeed one method of arriving at a set of stable weights, but it is also possible simply to report weights across studies and attempt to replicate results for "more" versus "less" severe events, roughly categorized. The latter approach, of course, assumes that the populations used are comparable. This does not advocate the use of specific weights per se; we concur with the weight of opinion that this is not a reasonable goal (Ross & Mirowsky, 1979). Rather, we believe that it is possible to find subsets of events that regularly surface across studies as the essential predictors. Following Shrout and others, we believe the point is more to identify and remove events from inventories that currently are in use but in fact are creating a "suppression" effect in correlations with health outcomes. In effect, this amounts to developing a cutoff for the importance of events—for example, its general statistical association across samples—and giving such events a weight of one, and others that rarely (if ever) have relevance a weight of zero.

The question is whether this approach will make any difference. The results using regression-based weighting are promising; that is, the differences in effects of events using this approach versus others is substantial. For example, Ross and Mirowsky report a correlation of .46 with distress symptoms compared to .19 using the now standard undesirable count score. In our own data, derived from two separate studies in Toronto, we have found that the traditional variance explained by the unit-weighted count method of approximately 10 percent increases to about 20 to 26 percent, using either regression weighting (i.e., stepwise selected individual events) or bivariate correlations as the selection mechanism. Similar results have been obtained in other studies using the same event selection approach (N. Lin, 1993, personal communication). The point here is not the significance of this increase; all are very significant at the .0001 level or beyond. The point is that we would never keep items in a measure if we knew that such items were acting as suppressors.

Despite the fact that the two inventories employed in our Toronto studies do not entirely overlap in item content and that the two distress measures used were different, there is some cross-validation of selection items: 85 percent of the events selected in one study appear in the list selected in the other study; and if we take the other study as the baseline, 75 percent of its selected events appear in the first study. These results suggest some potentially important possibilities. There may be a core of quite universally important events in the areas one might expect, such as deaths of intimates, relationship problems, financial crises, welfare dependency, assault victimization, and job demotions. Correspondingly, there may also be a large class of events, perhaps over 50 percent in most lists, that have little association with health outcomes for most people. If this is the case, this "dead weight" in life events scales needs to be specified and removed, and there is a need for cumulative evidence across studies about likely candidates. The problem posed by these dead-weight events is that they may significantly reduce correlations with outcomes, offering a new interpretation of the classic concern with the modest association of life events with health outcomes (Rabkin & Struening, 1976; Thoits, 1983).

The problem with the criticism that regression-based weighting results are not generalizable is that it prevents finding out whether in fact they are. If the literature had taken proper note of the Ross and Mirowsky (1979) findings, and of the suggestion of Shrout (1981) and Tausig (1982) that some events should be zero-weighted, the existing literature might already contain the required information.

If regression weighting is used, we suggest that results be reported separately for traditional and any constructed indices based on the sample at hand. Note that any regression-weighted index built from results in the same sample is likely to build in some overspecificity due to "capitalizing on chance." Even if some of the increase in impact between traditional unit-weighted counts and regression-based weighting is due to chance, the size of the increase suggests that there are real differences in impact captured by selection of subsets of events.

We need to be careful about two issues in removing events from lists. First, events may have only indirect effects on health outcomes through the occurrence of other consequent events. Thus, the fact that an event does not have a direct impact does not mean that an event is unimportant. Second, events can still have important conditional impacts in subpopulations of interest, even if their relevance in general populations is minimal. Thus, if one is studying the elderly, it is wise to retain "retirement," even if it has no impact in general populations.

Reliability Concerns

Reliability is a concern in that most studies report test–retest correlations for reports of events in the same period in the .4 to .7 range for test–retest intervals of 6 months or more (Neugebauer, 1981; Zimmerman, 1983). Some observers point out that the shorter time period is suspect, since the respondent may remember his or her earlier answer. Thus, the reliability as measured by test–retest correlations appears quite problematic. Some studies of reliability also assess the relevance of event characteristics, salience, and recency on consistency of reporting, and others use specific techniques during probing about events to increase recall accuracy (Brown & Harris, 1978).

We do not wish to dispute the fact that reliability of event reporting is a problem, for some types of events and lengths of recall. Rather, we question the use of test–retest correlations based on *total event scores*. This practice allows a score of 2, constituted by reporting a job loss and a divorce, to be consistent with a later of score of 2 after reporting an assault and a family member dying. The reliability of the total is important, but it is better to achieve that reliability by investigating the consistency of reporting individual items on inventories. Just as we suspect that events vary widely in their natural stress potential, we suspect that they also vary in their reliability. We are more likely to achieve greater reliability, if we pay attention to the differential reliability of items, rather than total scores.

A recent study by Kessler and Wethington (1991) offers a new approach and more hopeful evidence about the reliability issue. They reason that traditional approaches to measuring reliability, like test–retest correlations, share the common problem of taking all errors to be equally problematic. But in the case of life events, inconsistencies are likely to result predominantly from failure to report events that occur and less often from the reporting of fictitious events. Based on this reasoning, the authors develop a new measure of reliability that does not count the presence of a report by only one rater necessarily as an error. Rather, the reliability is estimated using the probability of either of two raters reporting an event compared to an *estimated* number of true events assuming independent reporting across raters, much as one can infer all cells in a 2 × 2 table, given information about some cells. In addition, Kessler and Wethington use a range of techniques to improve reliability overall, including (1) cues to improve memory; (2) wording questions in such a way as to define the realm of events; and (3) a life events calendar which helps date onset and offset of events. These efforts result in substantially higher reliabilities than are usually reported in the literature, often above .80 for important events. Of course, the logic of multiple informants applies to multiple reports from the same informant over time. This method depends essentially on the idea that underreporting is more typical than imagined events, an assumption that may hold more clearly for discrete, serious events within the last year than for childhood or more ongoing events without clear onset or offset.

Conclusions

Our major recommendations can be summarized as follows:

1. We advocate the specification of events as a separate realm, distinguished from other sources of and types of stress, so that the importance of different types of stress can be compared and distinguished from both prior circumstances and associated consequences.
2. We do not advocate the use of one specific published list over others; rather we think that many different variations of lists are roughly equivalent in their operation.
3. The tailoring of events to fit the risk status of populations is recommended, as long as there is a core of commonly important events that tend to supersede cultural differences as well.
4. The literature justifies exclusion of clearly positive events and suggests focus on negative events.
5. Events that are direct measurement indicators of physical or mental health status or of other forms of stress (i.e., chronic) need to be removed, but events that are only causally affected by health status should be retained.
6. A "middle-range" of from 30 to 50 items seems both optimal in terms of predictive power and efficient in terms of use of time and space.
7. We believe that problems due to lack of comprehensiveness can be corrected to some degree by the use of an open-ended follow-up to the regular inventory for the purpose of eliciting unusual events.
8. Variation in role occupancy should be examined and incorporated as a control in estimating life event effects.
9. A minimum of a 1-year time frame should be used in asking about events, with longer periods justified for more severe or the most objectively verifiable events.
10. Information about the time course of each event should be gathered;
11. We agree with the emphasis in the literature that unweighted indices are as useful as any, but we also encourage separate analysis of selected events based on associations with outcomes and the effect of removing events that act as suppressors.
12. We believe that the reliability issue is somewhat misspecified in the literature and support the investigation of the consistency of reporting for individual items, following the techniques suggested by Kessler and Wethington (1991).

We have assumed throughout that the study of life events is the study of only one form of stress exposure. Many forms of contextual factors may modify and shape the consequences of life events. It is important to realize that these factors that form the context of the occurrence of life events are not subservient in an overall model of health; they exist also as independent sources of health risk, quite apart from their implications for the impact of life events. It is possible to include "context" in a number of ways in studying the effect of events. Even beyond those already mentioned, it is important to note that the array of other players in the stress process— prior chronic stress, the state of preexisting coping resources, concomitant events within and across social roles, to name a few examples—should also be seen as providing this contextual information. The difference between this approach to specifying context and other approaches is that the contextual effects of these factors become part of the process of estimating the optimal model for prediction of health status, both as separate factors and as conditioners of the impact of life events.

Our approach to the application of life events involves the following irony: We advocate better totalities through careful attention to the individual parts. The tendency to treat the event total as an unassailable whole probably has had unfortunate consequences for the life event literature. At the same time, we do not believe that the traditional reliance on a summed score of life events—in some form—should be or will be supplanted by other approaches entirely. It is important to see that consideration of life events as a total score represents a specific piece of the stress argument, namely that there is a generic demand component to stressful occurrences that can be captured only by purposely adding up events despite content differences. In fact, before we "throw out the baby with the bath water," we need to remember that the implicit reason for counting and adding up stress exposure totals was exactly to specify the overall effects of demand, quite apart from the differences in content of stressors. Problems due to ignoring this content do suggest other approaches, such as the study of individual events, but, to turn the issue around, studying individual events does not capture the impact of total exposure either. This fact implies, we believe, that the effect of event totals can and should be considered as a background issue even in studies of the effects of individual events, such as divorce, widowhood, or job loss. This assumes, of course, that the specific event under study is removed from the remaining stress exposure total. Still, aggregate event stress exposure is meant to stand for the level of other sources of demand for adjustment apart from this event and, thus, is part of the contextual reality the individual faces in addition to the event of interest.

References

Aneshensel, C. S. (1992). Social stress: Theory and research. *Annual Review of Sociology, 18,* 15–38.

Aneshensel, C. S., & Frerichs, R. R. (1982). Stress, support and depression: A longitudinal causal model. *Journal of Community Psychology, 10,* 363–376.

Askenasy, A. R., Dohrenwend, B. P., & Dohrenwend, B. S. (1977). Some effects of social class and ethnic group membership on judgements of the magnitude of stressful life events: A research note. *Journal of Health and Social Behavior, 18,* 432–439.

Avison, W. R., & Turner, R. J. (1988). Stressful life events and depressive symptoms: Disaggregating the effects of acute stressors and chronic strains. *Journal of Health and Social Behavior, 29,* 253–264.

Barnett, B.E.W., Hanna, B., & Parker, G. (1983). Life event scales for obstetric groups. *Journal of Psychosomatic Research, 27* (4), 313–320.

Barrett, J. E., Rose, R. M., & Klerman, G. L. (Eds.), (1979). *Stress and mental disorder*. New York: Raven.

Billings, A. G., & Moos, R. H. (1982). Psychosocial theory and research on depression: An integrative framework and review. *Clinical Psychology Review, 2*, 213–237.

Birtchnell, J. (1988). Depression and family relationships: A study of young married women on a London housing estate. *British Journal of Psychiatry, 153*, 758–769.

Bramwell, S. T., Wagner, N. N., Masuda, M., & Holmes, T. H. (1975). Psychosocial factors in athletic injuries. *Journal of Human Stress, 1*, 6–20.

Brown, G. R., & Anderson, B. W. (1991). Psychiatric morbidity in adult inpatients with childhood histories of sexual and physical abuse. *American Journal of Psychiatry, 148*, 55–61.

Brown, G. W. (1981). Contextual measures of life events. In B. S. Dohrenwend & B. P. Dohrenwend (Eds.), *Stressful life events and their contexts* (pp. 187–201). New York: Prodist.

Brown, G. W. (1989). Life events and measurement. In G. W. Brown & T. O. Harris (Eds.), *Life events and illness* (pp. 3–45). New York: Guilford.

Brown, G. W., & Harris, T. O. (1978). *Social origins of depression*. New York: Free Press.

Bryer, J. B., Nelson, B. A., & Miller, J. B. (1987). Childhood sexual and physical abuse as factors in adult psychiatric illness. *American Journal of Psychiatry, 144*, 1426–1430.

Cleary, P. J. (1980). A checklist for life event research. *Journal of Psychosomatic Research, 24*, 199–207.

Cleary, P. J. (1981). Problems of internal consistency and scaling in life events schedules. *Journal of Psychosomatic Research, 25*(4), 309–320.

Cronbach, L. J. (1951). Coefficient alpha and the internal structure of tests. *Principles of Educational Psychological Measures, 18*, 132–165.

Dohrenwend, B. S. (1973). Life events as stressors: A methodological inquiry. *Journal of Health and Social Behavior, 14*, 167–175.

Dohrenwend, B. S., & Dohrenwend, B. P. (Eds.), (1974). *Stressful life events: Their nature and effect*. New York: Wiley.

Dohrenwend, B. S., & Dohrenwend, B. P. (1978). Some issues in research on stressful life events. *Journal of Nervous and Mental Disease, 166*, 7–15.

Dohrenwend, B. S., Krasnoff, L., Askenasy, A. R., & Dohrenwend, B. P. (1978). Exemplification of a method for scaling life events: The PERI life events scales. *Journal of Health and Social Behavior, 19*, 205–229.

Dohrenwend, B. S., Krasnoff, L., Askenasy, A. R., & Dohrenwend, B. P. (1982). The psychiatric epidemiology research interview life events scale. In L. Goldberger & S. Breznitz (Eds.), *Handbook of stress: Theoretical and clinical aspects*. New York: Free Press.

Dohrenwend, B. P., Raphael, K. G., Schwartz, S., Stueve, A., & Skodol, A. (1993). The structured event probe and narrative rating method for measuring stressful life events. In L. Goldberger & S. Breznitz (Eds.), *Handbook of stress* (pp. 174–199). New York: Free Press.

Eaton, W. W. (1978). Life events, social supports, and psychiatric symptoms: A re-analysis of the New Haven data. *Journal of Health and Social Behavior, 19*(2), 230–234.

Faravelli, C., Ambonetti, A., & Pallanti, S. (1986). Depressive relapses and incomplete recovery from index episode. *American Journal of Psychiatry, 143*, 888–891.

Grant, I., Sweetwood, H. L., Yager, J., & Gerst, M. S. (1981). Quality of life events in relations to psychiatric symptoms. *Archives of General Psychiatry, 38*, 335–339.

Hawkins, N. G., Davies, R., & Holmes, T. H. 1957. Evidence of psychosocial factors in the development of pulmonary tuberculosis. *American Review of Tuberculosis and Pulmonary Diseases, 75*, 768–780.

Holmes, T. H. (1979). Development and application of a quantitative measure of life change magnitude. In J. E. Barrett, R. M. Rose, & G. L. Klerman (Eds.), *Stress and mental disorder.* New York: Raven Press.

Holmes, T. H., & Masuda, M. (1972). Psychosomatic syndrome: When mothers-in-law or other disasters visit, a person can develop a bad, bad cold. Or worse. *Psychology Today,* 72–73.

Holmes, T. H., & Masuda, M. (1974). Life change and illness susceptibility. In B. S. Dohrenwend & B. P. Dohrenwend (Eds.), *Stressful life events: Their nature and effects* (pp. 45–72). New York: Wiley.

Holmes, T. H., & Rahe, R. H. (1967). The social readjustment rating scale. *Journal of Psychosomatic Research, 11,* 213–218.

Hough, R. L., Fairbank, D. T., & Garcia, A. M. (1976). Problems in the ratio measurement of life stress. *Journal of Health and Social Behavior, 17,* 70–82.

Hudgens, R. W. (1974). Personal catastrophe and depression: A consideration of the subject with respect to medically ill adolescents, and a requiem for retrospective life-event studies. In B. S. Dohrenwend & B. P. Dohrenwend (Eds.), *Stressful life events: Their nature and effects* (pp. 119–134). New York: Wiley.

Hurst, M. W. (1979). Life changes and psychiatric symptom development: Issues of context, scoring and clustering. In J. E. Barrett (Ed.), *Stress and mental disorder* (pp. 17–36). New York: Raven Press.

Jemmott, J. B., & Locke, S. E. (1984). Psychosocial factors, immunologic mediation, and human susceptibility to infectious diseases: How much do we know? *Psychological Bulletin, 95*(1), 78–108.

Jenkins, D. C. (1976). Recent evidence supporting psychologic and social risk factors for coronary disease. *New England Journal of Medicine, 294,* 987–994, 1033–1038.

Kahn, R., & Quinn, R. (1970). Role stress: A framework for analysis. In A. McLean (Ed.), *Mental health and work organizations* (pp. 50–115). Chicago: Rand McNally.

Kessler, R. C. (1983). Methodological issues in the study of psychosocial stress. In H. Kaplan (Ed.), *Psychosocial stress: Trends in theory and research* (pp. 267–341). Academic Press: New York.

Kessler, R. C., & Magee, W. J. (1993). Childhood adversities and adult depression: Basic patterns of association in a U.S. national survey. *Psychological Medicine, 23,* 679–690.

Kessler, R. C., Price, R. H., & Wortman, C. B. (1985). Social factors in psychopathology: Stress, social support, and coping processes. *Annual Review of Psychology, 36,* 531–572.

Kessler, R. C., & Wethington, E. (1991). The reliability of life event reports in a community survey. *Psychological Medicine, 21,* 723–738.

Lauer, R. H., & Lauer, J. C. (1991). The long-term relational consequences of problematic family backgrounds. *Family Relations, 40,* 286–290.

Liem, R., & Liem, J. (1978). Social class and mental illness reconsidered. The role of economic stress and social support. *Journal of Health and Social Behavior, 19,* 139–156.

Lloyd, C., Alexander, A. A., Rice, D. G., & Greenfield, N. S. (1980). Life events as predictors of academic performance. *Journal of Human Stress, 6,* 15–25.

Masuda, M., & Holmes, T. H. (1967a). Magnitude estimations of social readjustments. *Journal of Psychosomatic Research, 11,* 219–225.

Masuda, M., & Holmes, T. H. (1967b). The social readjustment rating scale: A cross cultural study of Japanese and Americans. *Journal of Psychosomatic Research, 11,* 227–237.

McFarlane, A. H., Norman, G. R., Streiner, D. L., Roy, R., & Scott, D. J. (1980). A longitudinal study of the influence of the psychosocial environment on health status: A preliminary report. *Journal of Health and Social Behavior, 21,* 124–133.

McLeod, J. (1991). Childhood parental loss and adult depression. *Journal of Health and Social Behavior, 35,* 205–220.

McQuaid, J. R., Monroe, S. M., Roberts, J. R., Johnson, S. L., Garamoni, G. L., Kupfer, D. J., & Frank E. (1992). Toward the standardization of life stress assessment: Definitional discrepancies and inconsistencies in methods. *Stress Medicine, 8,* 47–56.

Miller, T. W. (1989). Life-event scaling: Clinical methodological issues. In T. Miller (Ed.), *Stressful life events: International Universities Press Stress and Health Series, Monograph 4* (pp. 105–121). University Press.

Minter, R. E., & Kimble, C. P. (1978). Life events and illness onset: A review. *Psychosomatics, 19,* 334–339.

Monroe, S. M. (1982a). Life events assessment: Current practices, emerging trends. *Clinical Psychological Review, 2,* 435–453.

Monroe, S. M. (1982b). Assessment of life events: Retrospective versus concurrent strategies. *Archives of General Psychiatry, 39,* 606–610.

Monroe, S. M. (1982c). Life events and disorder: Event–symptom associations and the course of disorder. *Journal of Abnormal Psychology, 91*(1), 14–24.

Moos, R. H., & Swindle, R. W., Jr. (1990). Stressful life circumstances: Concepts and measures. *Stress Medicine, 6,* 171–178.

Mueller, D. P., Edwards, D. W., & Yarvis, R. M. (1977a). Stressful life events and psychiatric symptomatology: Change or undesirability. *Journal of Health and Social Behavior, 18,* 307–316.

Mueller, D. P., Edwards, D. W., & Yarvis, R. M. (1977b). Stressful life events and community mental health center patients. *Journal of Nervous and Mental Disease, 165,* 16–24.

Murphy, E., & Brown, G. W. (1980). Life events, psychiatric disturbance, and physical illness. *British Journal of Psychiatry, 136,* 326–338.

Myers, J. K., Lindenthal, J. J., & Pepper, M. P. (1974). Social class, life events, and psychiatric symptoms: A longitudinal study. In B. S. Dohrenwend & B. P. Dohrenwend (Eds.), *Stressful life events: Their nature and effects* (pp. 191–205). New York: Wiley.

Myers, J., Lindenthal, J., & Pepper, M. (1975). Life events, social integration and psychiatric symptomatology. *Journal of Health and Social Behavior, 14,* 412–427.

Neugebauer, R. (1981). The reliability of life events reports. In B. S. Dohrenwend and B. P. Dohrenwend (Eds.), *Stressful life events and their context* (pp. 85–107). New York: Prodist.

Newcomb, M. D., Huba, G. J., & Bentler, P. M. (1986). Life change events among adolescents: An empirical consideration of some methodological issues. *Journal of Nervous and Mental Disease, 174*(5), 280–289.

Parker, G., & Hadzi-Pavlovic, D. (1984). Modification of levels of depression in mother-bereaved women by parental and marital relationships. *Psychological Medicine, 14,* 125–135.

Paykel, E. S. (1978). Contribution of life events to causation of psychiatric illness. *Psychological Medicine, 8,* 245–253.

Paykel, E. S. (1979). Causal relationships between clinical depression and life events. In J. E. Barrett (Ed.), *Stress and mental disorder* (pp. 71–86). New York: Raven Press.

Paykel, E. S., Prusoff, B. A., & Uhlenhuth, E. H. (1971). Scaling of life events. *Archives of General Psychiatry, 25,* 340–347.

Pearlin, L. I. (1983). Role strains and personal stress. In H. B. Kaplan (Ed.), *Psychosocial stress: Trends in theory and research* (pp. 3–31), New York: Academic Press.

Pearlin, L. I. (1989). The sociology study of stress. *Journal of Health and Social Behavior,* *30,* 241–256.

Pearlin, L. I., & Lieberman, M. A. (1979). Social sources of emotional distress. In R. G. Simmons (Ed.), *Research in community and mental health 1,* (pp. 217–248). Greenwich, CN: JAI.

Pearlin, L. I., Lieberman, M. A., Menaghan, E. G., & Mullan, J. T. (1981). The stress process. *Journal of Health and Social Behavior, 22,* 337–356.

Pugh, W., Erickson, J. M., Rubin, R. T., Gunderson, E.K.E., & Rahe, R. H. (1971). Cluster analyses of life changes. II. Method and replication in Navy subpopulations. *Archives of General Psychiatry, 25,* 333–339.

Rabkin, J. G., & Struening, E. L. (1976). Life events, stress and illness. *Science, 194,* 1013–1020.

Raphael, E. G., Cloitre, M., & Dohrenwend, B. P. (1991). Problems of recall and misclassification with checklist methods of measuring stressful life events. *Health Psychology, 10,* 62–74.

Ross, C. E., & Huber, J. (1985). Hardship and depression. *Journal of Health and Social Behavior, 26*(4), 312–327.

Ross, C. E., & Mirowsky, J. (1979). A comparison of life-event weighting schemes: Change, undesirability, and effect-proportional indices. *Journal of Health and Social Behavior, 20,* 166–177.

Rutter, M. (1989). Pathways from childhood to adult life. *Journal of Child Psychology and Psychiatry, 30,* 23–51.

Sandler, I. N., & Block, M. (1979). Life stress and maladaptation of children. *American Journal of Community Psychology, 7,* 425–440.

Sandler, I. S., & Guenther, R. T. (1985). Assessment of life stress events. In P. Karoly (Ed.), *Measurement strategies in health psychology* (pp. 555–600). New York: Wiley.

Schroeder, D. H., & Costa, P. T. (1984). Influence of life event stress on physical illness: Substantive effects or methodological flaws? *Journal of Personality and Social Psychology, 46,* 853–863.

Selye, H. (1956). *The stress of life.* New York: McGraw-Hill.

Selye, H. (1976). *The stress of life* (rev. ed.). New York: McGraw-Hill.

Selzer, M. L., & Vinokur, A. (1974). Life events, subjective stress, and traffic accidents. *American Journal of Psychiatry, 131,* 903–906.

Shrout, P. E. (1981). Scaling of stressful life events. In B. S. Dohrenwend & B. P. Dohrenwend (Eds.), *Stressful life events and their contexts* (pp. 29–47). New York: Prodist.

Swearingen, E. M., & Cohen, L. H. (1985). Measurement of adolescents' life events: The junior high life experiences survey. *American Journal of Community Psychology, 13,* 69–85.

Tausig, M. (1982). Measuring life events. *Journal of Health and Social Behavior, 23,* 52–64.

Tennant, C. (1988). Parental loss in childhood: Its effect in adult life. *Archives of General Psychiatry, 45,* 1045–1050.

Thoits, P. A. (1981). Undesirable life events and psychophysiological distress: A problem of operational confounding. *American Sociological Review, 46*(1), 97–109.

Thoits, P. A. (1983). Dimensions of life events that influence psychological distress: An evaluation and synthesis of the literature. In H. B. Kaplan (Ed.), *Psychosocial stress: Trends in theory and research* (pp. 33–101). New York: Academic Press.

Thoits, P. A. (1987). Gender and marital status differences in control and distress: Common stress versus unique stress explanations. *Journal of Health and Social Behavior 28*(1), 7–22.

Thurlow, H. J. (1971). Illness in relation to life situation and sick role tendency. *Journal of Psychosomatic Research, 15,* 73–88.

Turner, R. J., & Avison, W. R. (1992). Innovations in the measurement of life stress: Crisis theory and the significance of event resolution. *Journal of Health and Social Behavior, 33,* 36–50.

Turner, R. J., & Noh, S. (1988). Physical disability and depression: A longitudinal analysis. *Journal of Health and Social Behavior, 29,* 263–277.

Turner, R. J., & Roxburgh, S. J. (in press). Psychological adjustment and the mothering role: A longitudinal assessment of the significance of life stress and social support. In P. J. Leaf (Ed.), *Research in community and mental health.*

Turner, R. J., Wheaton, B., & Lloyd, D. (in press) The epidemiology of social stress. *American Sociological Review.*

Vinokur, A., & Selzer, M. L. (1975). Desirable versus undesirable life events: Their relationship to stress and mental distress. *Journal of Personality & Social Psychology, 32,* 329–337.

Wheaton, B. (1991). The specification of chronic stress: Models and measurement. Presented at the *Society for the Study of Social Problems Annual Meetings,* August, 1991.

Wyler, A., Masuda, M., & Holmes, T. H. (1971). Magnitude of life events and seriousness of illness. *Psychosomatic Medicine, 33,* 115–120.

Zautra, A. J., & Reich, J. W. (1983). Life events and perceptions of life quality: Developments in a two-factor approach. *Journal of Community Psychology, 1,* 121–132.

Zimmerman, M. (1983). Methodological issues in the assessment of life events: A review of issues and research. *Clinical Psychology Review, 3,* 339–370.

Appendix 1

Life Event Inventories

1. Children

Childhood Life Events and Family Characteristics Questionnaire. Byrne, C. P., Velamoor, V. R., Cernovsky, Z. Z., Cortese, L., & Losztyn, S. (1990). A comparison of borderline and schizophrenic patients for childhood life events and parent–child relationships. *Canadian Journal of Psychiatry, 35*(7), 590–595.

Children's Life Event Questionnaire, 1989. Deutsch, L. J., and Erickson, M. T. (1989). Early life events as discriminators of socialized and undersocialized delinquents. *Journal of Abnormal Child Psychology, 17*(5), 541–551.

Divorce Events Schedule for Children. Sandler, I. N., Ramirez, R., & Reynolds, K. D. "Life stress for children of divorce, bereaved, and asthmatic children. Paper presented at the American Psychological Association convention, Washington D.C., August 1986. Referenced in Roosa, M. W., Beals, J., Sandler, I. N., & Pillow, D. R. (1990). The role of risk and protective factors in predicting symptomatology in adolescent and self-identfied of alcoholic parents. *American Journal of Community Psychology, 18*(5): 25–741.

General Life Events Schedule for Children. Sandler, I. N., Nolichuk, S., Brauer, S. L., & Fogas, B. (1986). Significant events of children of divorce: Toward the assessment of a risky situation. In S. M. Averback & A. Stolberg (Eds.), *Crisis intervention with children and families* (pp. 65–87). New York: Hemisphere.

The Children of Alcoholics Life-Events Schedule. Roosa, M. W., Sandler, I. N., Gehring, M., Beals, J. & Cappo, L. (1988). The children of alcoholics life-events schedule: A stress scale for children of alcohol abusing parents. *Journal of Studies on Alcohol, 49*(5), 422–429.

Yamamoto, K. (1979). Children's ratings of the stressfulness of experiences. *Developmental Psychology, 15,* 581–582.

2. Adolescents

Adolescent–Family Inventory of Life Event Changes. McCubbin, H. I., Patterson, Bauman, E., & Harris, L. (1981). Adolescent–family inventory of life event changes. Madison, Wis.: University of Wisconsin. Referenced in Carty, L. (1989). Social support, peer counselling, and the community counsellor. *Canadian Journal of Counselling, 23*(1), 92–102.

Adolescent Life Change Event Scale. Hussey M., & Ingle, M. (1977). The development of a life change event scale for adolescents. Unpublished Master's thesis, University of Cincinnati. Modified by Yeaworth, R., York, J., Hussey, M., Ingle, M., & Goodwin, T. 1980. The development of an adolescent life changes event scale. *Adolescence, 15,* 91–97.

Adolescent Life Experiences Survey. Towbes, L. C., Cohen, L. H., & Glyshaw, K. (1989). Instrumentality as a life-stress moderator for early versus middle adolescents. *Journal of Personality and Social Psychology, 57*(1), 109–119.

Adolescent Perceived Events Scale. Compas, B. E., Davis, G. E., Forsythe, C. G., & Wagner, B. M. (1987). Assessment of major and daily life events during adolescence: The Adolescent Perceived Events Scale. *Journal of Consulting and Clinical Psychology, 55,* 534–541.

Burke, R. J., & Weir, T. (1978). Sex differences in adolescent life stress, social support, and well-being. *Journal of Psychology, 98,* 277–288.

Coddington, R. D. (1972a). The significance of life events as etiologic factors in the diseases of children–I. *Journal of Psychosomatic Research, 16,* 7–18. Also Coddington, R. D. (1972b). The significance of life events as etiologic factors in the diseases of children–II. *Journal of Psychosomatic Research, 16,* 205–213.

Coddington, R. D. (1979). Life events associated with adolescent pregnancies. *Journal of Clinical Psychiatry, 40,* 180–185.

Junior High Life Events Survey. Swearingen, E. M., & Cohen, L. H. (1985). Life events and psychological distress: A prospective study of young adolescents. *Developmental Psychology, 21,* 1045–1054.

Gad, M. T., & Johnson, J. H. (1980). Correlates of adolescent life stress as related to race, SES and levels of perceived social support. *Journal of Clinical Child Psychology* (Spring), 13–16. Modification of Life Experiences Survey by Sarason et al. (1978).

Gersten, J. C., Langner, T. S., Eisenberg, J. G., & Simcha-Fagan, O. (1977). An evaluation of the etiologic role of stressful life-change events in psychological disorders. *Journal of Health and Social Behavior, 18,* 228–244.

High School Social Readjustment Scale. Tolor, A., Murphy, V., Wilson, L. T., & Clayton, J. (1983). The High School Readjustment Scale: An attempt to quantify stressful events in young people. *Research Communication in Psychology, Psychiatry and Behavior, 8,* 85–111.

Kaplan, H. B., Robbins, C., & Martin, S. S. (1983). Antecedents of psychological distress in young adults: self-rejection, deprivation of social support, and life events. *Journal of Health and Social Behavior, 24:* 230–244.

Life Events Checklist. Johnson, J. H., & McCutcheon, S. M. (1980). Assessing life stress in older children and adolescents: Preliminary findings with the life events checklist. In I. G. Sarason & C. C. Spielberger (Eds.), *Stress and anxiety* (Vol. 7, pp. 111–125). Washington, D.C.: Hemisphere.

Life Event Scale for Adolescents. Coddington, R. D., & Troxell, J. R. (1980). The effect of emotional factors on football injury rates: A pilot study. *Journal of Human Stress, 6,* 3–5.

Life Stress Inventory. Cohen-Sandler, R., Berman, A. L., & King, R. A. (1982). Life stress and symptomatology: Determinants of suicidal behavior in children. *Journal of the American Academy of Child Psychiatry, 21,* 178–186.

Newcomb, M. D., Huba, G. J., & Bentler, P. M. (1981). A multidimensional assessment of stressful life events among adolescents: Derivation and correlates. *Journal of Health and Social Behavior, 22,* 400–415.

Warrick Life Events Checklist. Warrick L. (1987). Evaluation to Flinn Foundation school leased health care program for pregnant and parenting teens, 1986–1989. Tucson, AZ: Flinn Foundation.

Wright, L. S. (1985). Suicidal thoughts and their relationship to family stress and personal problems among high school seniors and college undergraduates. *Adolescence, 20*(79), 575–580.

3. Adults

Barnett, B.E.W., Hanna, B., & Parker, G. (1983). Life event scales for obstetric groups. *Journal of Psychosomatic Research, 27*(4), 313–320.

Block, M., & Zautra, A. (1981). Satisfaction and distress in a community: A test of the effects of life events. *American Journal of Community Psychology, 9*, 165–180.

Hough, R. L., Fairbank, D. T., & Garcia, A. M. (1976). Problems in the ratio measurement of life stress. *Journal of Health and Social Behavior, 17*, 70–82.

Interview Schedule for Events and Difficulties. Brown, G. W., & Harris, T. O. (1978). *Social origins of depression.* New York: The Free Press.

Junior High Life Experiences Survey. Swearingen, E. M., & Cohen, L. H. (1985). Measurement of adolescents' life events: The junior high life experiences survey. *American Journal of Community Psychology, 13*, 69–85.

Life Change Inventory. Costantini, A. F., Braun, J. R., Davis, J., & Iervolino, A. (1974). The life change inventory: A device for quantifying psychological magnitude of changes experienced by college students. *Psychological Reports, 34*, 991–1000.

Life Crisis History. Antonovsky, A., & Kats, R. (1967). The life crisis history as a tool in epidemiological research. *Journal of Health and Social Behavior, 8*, 15–21.

Life Events Inventory. Cochrane, R., & Robertson, A. (1973). The life events inventory: A measure of the relative severity of psychosocial stressors. *Journal of Psychosomatic Research, 17*, 135–139.

Life Events Inventory (Questionnaire Form). Tennant, C., & Andrews, G. (1976). A scale to measure the stress of life events. *Australian and New Zealand Journal of Psychiatry, 10*, 27–32.

Live Events Questionnaire (Short and Long Forms). Horowitz, M. J., Schaefer, C., Hiroto, D., Wilner, N., & Levin, B. (1977). Life event questionnaires for measuring presumptive stress. *Psychosomatic Medicine, 39*, 413–431.

Life Events Questionnaire–II. Cooley, E. J., Miller, A. W., Keesey, J. C., Levenspiel, M. J., & Sisson, C. F. (1979). Self-Report assessment of life change and disorders. *Psychological Reports, 44*, 1078–1086.

Life Experiences Survey. Sarason, I. G., Johnson, J. H., & Siegel, J. M. (1978). Assessing the impact of life changes: Development of the Life Experiences Survey. *Journal of Consulting and Clinical Psychology, 46*, 32–46.

Paykel, E. S., Pursoff, B. A., & Uhlenhuth, E. H. (1971). Scaling of life events. *Archives of General Psychiatry, 25*, 340–347.

Psychiatric Epidemiology Research Interview for Life Events. Dohrenwend, B. S., Krasnoff, L., Askenasy, A. R., & Dohrenwend, B. P. (1978). Exemplification of a method for scaling life events: The PERI Life Events Scale. *Journal of Health and Social Behavior, 19*, 205–229.

Recent Life Change Questionnaire. Rahe, R. H. (1975). Epidemiological studies of life change and illness. *International Journal of Psychiatry in Medicine, 6:* 133–146.

Review of Life Events. Hurst, M. W., Jenkins, C. D., & Rose, R. M. (1978). The assessment of life change stress: A comparative and methodological inquiry. *Psychosomatic Medicine, 40*, 126–141.

Social Readjustment Rating Scale. Holmes, T. H., & Rahe, R. H. (1967). The Social Readjustment Rating Scale. *Journal of Psychosomatic Research, 11*, 213–218.

Social Readjustment Scale for Children. Coddington, R. D. (1972a). The significance of life events as etiologic factors in the diseases of children I. A survey of professional workers. *Journal of Psychosomatic Research, 16*, 7–18.

Social Readjustment Scale for Children. Coddington, R. D. (1972b). The significance of life events as etiologic factors in the diseases of children II. A survey of a normal population. *Journal of Psychosomatic Research, 16*, 205–213.

Structured Event Probe and Narrative Rating Method. Dohrenwend, B. P., Raphael, K. G., Schwartz, S., Stueve, A., & Skodol, A. (1993). The structured event probe and narrative rating method for measuring stressful life events. In L. Goldberger & S. Brenitz (Eds.), *Handbook of stress* (pp. 174–199). New York: Free Press.

4. Aged

Geriatric Social Readjustment Rating Scale. Amster L., & Krauss, H. (1974). The relationship between life crises and mental deterioration in old age. *International Journal of Aging and Human Development, 5,* 51–55.

Inventory of Small Life Events (modified for the elderly). Zautra, A. J., Guarnaccia, C. A., & Dohrenwend, B. P. (1986). Measuring small life events. *American Journal of Community Psychology, 14,* 629–655.

Louisville Older Person Event Scale. Murrell, S. A., Norris, F. H., & Hutchins, G. M. (1984). Distribution and desirability of life events in older adults: Population and policy implications. *Journal of Community Psychology, 12,* 301–311.

Appendix 2

Sampling of Critical Reviews of Stressful Life Event Methodology

Brown, G. W., & Harris, T. (1978). *Social origins of depression.* New York: Free Press.

Cleary, P. A. (1981). Problems of internal consistency and scaling in life event schedules. *Journal of Psychosomatic Research, 25,* 309–320.

Dohrenwend, B. S., & Dohrenwend, B. P. (1978). Some issues in research on stressful life events. *Journal of Nervous and Mental Disease, 166,* 7–15.

Hurst, M. W. (1979). Life changes and psychiatric symptom development: Issues of context, scoring and clustering. In J. E. Barrett (Ed.), *Stress and mental disorder,* (pp. 17–36). New York: Raven Press.

McQuaid, J. R., Monroe, S. M., Roberts, J. R., Johnson, S. L., Garamoni, G. L., Jupfer, D. J., & Frank, E. (1992). Toward the standardization of life stress assessment: Definitional discrepancies and inconsistencies in methods. *Stress Medicine, 8,* 47–56.

Miller, T. W. (1989). Life events scaling: Clinical methodological issues. In T. Miller (Ed.), *Stressful life events. International Universities Press Stress and Health Series, Monograph 4,* (pp. 105–121). Universities Press.

Monroe, S. M. (1982). Life events assessment: Current practices, emerging trends. *Clinical Psychology Review, 2,* 435–453.

Moos, R. H., & Swindle, R. W., Jr. (1990). Stressful life circumstances: Concepts and measures. *Stress Medicine, 6,* 171–178.

Neugebauer, R. (1981). The reliability of life event reports. In B. S. Dohrenwend & B. P. Dohrenwend (Eds.), *Stressful life events and their context* (pp. 85–107). New York: Prodist.

Newcomb, M. D., Huba, G. J., & Bentler, P. M. (1986). Life change events among adolescents: An empirical consideration of some methodological issues. *Journal of Nervous and Mental Disease, 174*(5), 280–289.

Paykel, E. S. (1983). Methodological aspects of life events research. *Journal of Psychosomatic Research, 27,* 341–352.

Perkins, D. V. (1982). The Assessment of stress using life event scales. In L. Goldberger & S. Breznits (Eds.), *Handbook of stress: Theoretical and clinical aspects* (pp. 320–332). New York: Free Press.

Rabkin, J. G., & Struening, E. L. (1976). Life events, stress and illness. *Science, 194,* 1013–1020.

Raphael, E. G., Cloitre, M., & Dohrenwend, B. P. (1991). Problems of recall and misclassification with checklist methods of measuring stressful life events. *Health Psychology, 10,* 62–74.

Sandler, I. N. (1979). Life stress events and community psychology. In I. G. Sarason & C. D. Spielberger (Eds.), *Stress and anxiety* (Vol. 6, pp. 213–232). New York: Hemisphere.

Sarason, I. G., deMonchaux, C., & Hunt, T. (1975). Methodological issues in the assessment of life stress. In L. Levy (Ed.), *Emotions—Their parameters and measurement* (pp. 399–509). New York: Raven Press.

Tausig, M. (1982). Measuring life events. *Journal of Health and Social Behavior, 23,* 52–64.

Tennant, C., Bebbington, P., & Hurry, J. (1981). The role of life events in depressive illness: Is there a substantial causal relation? *Psychological Medicine, 11,* 379–389.

Thoits, P. A. (1983). Dimensions of life events as influences upon the genesis of psychological distress and associated conditions: An evaluation and synthesis of the literature. In Howard B. Kaplan (Ed.), *Psychosocial stress: Trends in theory and research* (pp. 33–103). New York: Academic Press.

Zimmerman, M. (1989). Methodological issues in the assessment of life events: A review of issues and research. *Clinical Psychology Review, 3,* 339–370.

3

Interview Measurement
of Stressful Life Events

Elaine Wethington, George W. Brown, and, Ronald C. Kessler

A critical component of establishing the overall relationship between environmental stressors and individual health and well-being is the utilization of a comprehensive, reliable, and valid measure of stressful life events and difficulties. Two methods of life events measurement predominate: checklist measures (see Chapter 2, this volume) and intensive personal interview measures, which use qualitative probes in order to specify more precisely the characteristics of life events believed to produce stress. These two sorts of measures have evolved from different, although not mutually exclusive, theories of what constitutes stress. As a concomitant to their origins in different theories of stress, these measures have also been shaped in response to different research questions and paradigms.

Checklist methods were informed by a theoretical perspective on stress which asserts that the magnitude of social and environmental change bringing about a need for readjustment is the basis of experienced stress (see Chapter 2, this volume, for a review). The early development of personal interview methods that use qualitative probes, however, was driven by a perspective which assumed that social and environmental changes (and anticipations of those changes) which threaten the most strongly held emotional commitments are the basis for experienced *severe* stress, and that it is severe stress which threatens health (Gorman & Brown, 1992). Other interview measures have since been developed that are based on the change–readjustment paradigm (e.g., Dohrenwend, Raphael, Schwartz, Stueve, & Skodol, 1993); these are all distinguished from the checklists, however, by their use of qualitative probes.

There are other critical differences between checklist and personal interview measures of life events. The life change–readjustment paradigm assumes that positive as well as negative changes are related to the experience of emotional stress; interview methods, however, consistent with their focus on the most emotionally arousing stressors, are designed to elicit reports of negative events. Checklist methods yield a summary score of the stressfulness of changes experienced over a period of time (usually a year), whereas interview methods are designed to elicit reports of specific events that may have triggered an onset of a severe physical or mental illness. In regard to this latter point, checklist measures tend to (but not solely) use continuous symptom or "mood" measures as outcomes. Interview measures tend to utilize as outcomes, onsets of clinically significant mental disorder or physical illness—for example, depression or gastrointestinal attacks.

The aim of this chapter is to acquaint readers with the circumstances in which the use of personal interview methods is appropriate; the relative strengths and weaknesses of available personal interview methods; and the logistics and costs of fielding an interview study. It is not the intent of this chapter to argue that personal interview methods are necessary in all cases for measuring exposure to stressors. Personal interview methods are not efficient for correlational or prospective studies, where the purpose is to establish that there is a relationship between environmental stress and a *general* measure of symptoms. In such instances, the greater expense of using interview measures is not justified. However, there are instances in which the use of personal interview methods is more suited to a particular research question.

Rationale for the Personal Interview Versus Checklist Approach

The use of the intensive personal interview technique is particularly critical when the relative timing of exposure to a stressor and the onset of an illness is necessary to address a research question. For example, if an investigator is interested in the relationship between experiencing a stressor and the subsequent onset of panic disorder, then both the timing of the events and the onset of panic must be carefully (and independently) dated. Precise dating is necessary not only to establish the relationship of stress exposure to onset, but also to identify the aspects of the stressful experience that affect onset. In general, self-administered checklists cannot equal the precision of an interviewer trained to elicit aspects of events critical to examine these research questions (Cannell, Miller, & Oksenberg, 1981). Studies of illness onset, course, and recovery often require such precise dating of event and illness onset/offset.

Checklist methods also seem to be more prone to "telescoping," or the misdating of distant events into a more recent time period (McQuaid et al., 1992; Raphael, Cloitre, & Dohrenwend, 1991). Checklist methods, moreover, are probably not as effective as more intensive personal interviewing in communicating the importance of accurate answers to the respondents (Cannell et al., 1981). In addition, the face-to-face interview with qualitative probes facilitates the use of memory aids to improve event recall and accurate dating (such as calendars, visual representations of important events of the preceding year, and well-timed reminders of personally salient dates). Such aids have been shown to provide substantially better information than can be collected through relatively unassisted self-report (Sobell, Toneatto, Sobell, Schuller, & Maxwell, 1990). This more accurate timing of event onset and offset, moreover, makes it possible to distinguish and date a series of related events and difficulties, which may be required in a study addressing whether exposure to multiple events is more related to onset than exposure to a single event (Brown & Harris, 1978; McQuaid et al., 1992; Raphael et al., 1991).

Interview measures of life events, however, are most widely promoted for their methods of establishing more precisely the degree of *severity* in stressor exposure.

Studies of responses to checklist measures have found that many respondents report relatively minor events in response to questions that are designed to elicit only severely stressful events (Dohrenwend, Link, Kern, Shrout, & Markowitz, 1990), such as reporting the death of someone relatively distant as the death of someone "close"; or reporting a broken toe as a serious injury, even when it had no impact on one's life besides short-term pain and inconvenience. In experiments with interview measures of life events, Kessler and Wethington (1991) have found that many people report objectively trivial events in response to questions that were thought to emphasize more than adequately the serious nature of the stressful events that were sought. Reporting minor events as severe, moreover, may be related to the respondent's health status at the time of interview (Bebbington, 1986): Distressed people are more likely to remember negative events than happy people (Fiedler & Stroehm, 1986).

Interview measures of life events, moreover, use a different system for rating the severity of life events. Checklist measures have generally (although not universally) applied "normative life change" or "readjustment" weights—a numerical value preassigned to every event reported or the sum of the number of events experienced over a given period (see Chapter 2). If the event description is relatively more inclusive (e.g., "serious illness"), the life change unit weight may be a poor fit to the severity of the actual event prompting the report. Specifically, a recurrence of cancer arguably has a different impact than a bout of influenza, both of which could prompt a yes answer to "serious illness." Studies comparing checklist to interview methods have confirmed that variability of this sort occurs when subjects respond to checklists. Raphael et al. (1991) and McQuaid et al. (1992) demonstrated that respondents to checklists report experiences to life event questions that do not match the researcher's intent. Furthermore, respondents are known to report minor or even positive events in response to questions that were designed to elicit only highly negative and undesirable events (McQuaid et al., 1992).

Event list comprehensiveness has also been a critical issue in the debate between the use of checklist and personal interview methods. Self-report event lists have been criticized for lacking coverage of singular and rare events that may be among the most distressing (Brown, 1981), although more comprehensive checklist methods have been developed in response to this criticism (Zimmerman, 1983). Personal interview methods allow for more flexibility in event elicitation and recording, as well as self-nomination of stressors not evoked by direct questions about domain-specific events and transitions (Brown & Harris, 1978; Dohrenwend et al., 1990).

Interview measures of life events also allow for more effective probing of event reports, an important means of reducing measurement error. An interviewer trained in study procedures can probe to determine if the event reported indeed "matches" the intent of the question and should be counted as a stressor of that type (McQuaid et al., 1992; Raphael et al. 1991). In addition, probing of event responses also allows investigators to determine whether events are connected to one another or whether the same event is mentioned more than once in response to separate questions (Kessler & Wethington, 1991; McQuaid et al., 1992). The latter is useful in order to counter the tendency of some respondents to "overreport" their exposure by mentioning the same event more than once.

Finally, interview methods have yielded some *different* insights into a number of critical issues in stress research, primarily because trained interviewers can elicit more detailed descriptive information about variations in event severity over time, and about linkages among related events and difficulties. Interview measures have been used to explore (1) the relationship among alleviation of environmental stressors, illness course, duration, and recovery (e.g., Brown, Adler, & Bifulco, 1988); (2) the sorts of stressors that are associated with different types of disease onset (e.g., Bebbington et al., 1988; Harris, 1991); (3) and the interaction of individual vulnerability and life history with recent stressors to produce onset of illness (Brown, Bifulco, & Harris, 1987).

Applications of Interview Measures of Life Events

Interview measures of life events are extremely expensive in comparison to check-list methods. Given their expense, the use of interview methods for measuring life events is most appropriate for field studies or retrospective or longitudinal case-control studies where the relationship of stress exposure to the *onset* of very serious illness is being investigated (e.g., Katschnig, 1986), and where exposure to life events is of primary interest to the researcher (Cooke, 1985). In such studies, the major research questions entail establishing the relationship of environmental stressors to mental or physical illness in the community. In these studies as well, a maximally comprehensive measure of stress exposure is necessary. The reason is that the focus in such studies is not on one particular event or transition that many people experience, and explaining variations in people's responsiveness to it, but on establishing the etiology of particular disorders or illnesses, particularly the role of stressful events.

Although personal interviewing *methods* have also been used extensively in studies of adjustment to one stressor, such as divorce, unexpected unemployment, or widowhood, this chapter will not review such studies, because our emphasis here is on comprehensive measures of stressor exposure. The intention is not to urge, even inadvertently, that researchers abandon intensive studies of one crisis in favor of a comprehensive stressor measurement approach. In a specific crisis study, the comprehensive stressor measures described here would be poor substitutes for the sort of fine-grained examinations of individual variation and adaptation that are typical in such studies.

Choosing an Appropriate Measure

The following section critically evaluates a number of interview measures of life events. The review is confined to those measures that have been used more than once, or that are readily available for replication by researchers other than their developers. The criteria for evaluation include (1) reliability (if reported); (2) comprehensiveness of event coverage; (3) falloff in reports of events over a 1-year event recall period; (4) the relationship of stressor exposure, as estimated by this interview method, to onset of illness; (5) the dimensions by which stressors are rated in order to evaluate their "stressfulness"; (6) illnesses and disorders to which the measure has been applied; (7) populations in which the measure has been used, including cultural groups; (8) strengths of the method and its potential for future development; and (9) major criticisms that have emerged. For several of the reviewed measures, not all of this information is available.

Interview methods are presented as follows: Because of its earlier development, the Brown and Harris Life Events and Difficulties Schedule (LEDS) is described first. It is followed by reviews of a number of interview life-event rating measures that do not reproduce the event rating system used in the LEDS (e.g., the SEP-RATE: see below and Dohrenwend et al., 1993). Finally, interview measures that are shortened or more structured versions of the LEDS are reviewed, including a structured form of the LEDS under development (the SLI: see below and Wethington, Kessler, & Brown, 1993) and interviews that use checklists to reduce interviewing time.

The Life Events and Difficulties Schedule

The most widely used personal interview method is the Life Events and Difficulties Schedule (LEDS: Brown & Harris, 1978). The LEDS is a semi-structured survey instrument, appropriate for use in a community sample as well as with patients, assessing a wide variety of stressors. The interview consists of a series of questions asking whether certain types of events had occurred over the past 12 months (or longer) and a set of guidelines for probing positive responses. Interviewers are given a great deal of latitude in deciding whether more intense probing is required for ascertaining whether an event is likely to have been severe in emotional impact, or not. To that end, only general guidelines (brief reminders about important objective dimensions) are given for probing, since the interview is designed to resemble conversation rather than a standard survey.

The aim of the interviewing method is to construct a narrative or "story" of each event (Gorman & Brown, 1992). Once an event is mentioned, the interviewer asks questions about objective circumstances surrounding the event, such as: What led up to the event? What followed it? Interviewers also follow up on spontaneous comments made by the respondents, including mentions of other events occurring at the same time, or what may be a technically "wrong" answer—for example, a parent's sudden death reported in response to a question about serious illness of a family member.

The purpose of the probes is to gather enough information to rate the *long-term contextual threat* of discrete events (a death) or *severity* of chronic difficulties (several months of unemployment). The rating of long-term contextual threat is the key component of the LEDS method, as the experience of a "severely" threatening event or presence of a severe chronic difficulty ("marked" in LEDS terminology) is hypothesized to pose a risk for the onset of physical or mental illness. Rating the degree of threat for specific sorts of events and difficulties has been developed and documented over the several decades in which the LEDS has been used. The documentation for the ratings is contained in *dictionaries* produced by Brown,

Harris, and their colleagues.[1] These dictionaries, which are made available to those who are formally trained in the use of the method, consist of several thousand examples of rated life events and difficulties, grouped by life domain (marital, work-related, financial), type of event or difficulty (argument, separation, ongoing marital problems), and level of rated severity (4 categories for events, and 7 for difficulties). In addition to the examples, the dictionaries also document how events are classified by type and the sorts of information the interviewer needs to establish in order to assign contextual threat ratings to each type of event.

The critical element of rating long-term contextual threat is collecting objective information in order to estimate the long-term implications of an event or difficulty for important *plans, concerns, and purposes* of the respondent, taking into account other current circumstances and relevant life history information. Aspects of life history and context that are relevant are documented in the life event and difficulty dictionaries. The general rule guiding event severity ratings is that the interviewer must determine whether the event affected the life plans or commitments of the respondent, and use the dictionary guidelines to assess those aspects of context that must be established in order to estimate the level of emotional arousal likely to have resulted. In many instances, however, the interviewer can rely on the event or difficulty having a specific rating in the LEDS dictionaries.

The event list in the LEDS consists not only of questions about probable negative events, but also of role transitions (births, promotions, graduation from school) and situations that may have involved a severe emotional reaction, but are not typically defined as events in checklists. An unusual feature of the LEDS are questions such as dealing with important revelations about other people's character, breaking or getting bad news, making difficult decisions, disappointments, and finding presumptive negative events working out much better than expected.

There are three main stages of life-event rating: (1) assessment of the likely emotional arousal (positive or negative) associated with the reported event; (2) a rating of general contextual threat, which has a very high threshold for the highest rating of severity; and (3) ratings of specific aspects of threat, such as its degree of loss (of a person, object, or idea), danger (the future probability of loss), frustration of goals, challenge, intrusiveness into daily routine, and positiveness (e.g., relief, or getting a fresh start on life). Events are rated for both their positive and negative aspects. Difficulties are rated for their intrusiveness into daily routine and their threat to the quality of other life roles, such as work.

The interviews average about one hour to complete in a general population sample, with a range from 30 minutes to well over 2 hours. After the interviewer completes the interview, he or she then constructs written narrative descriptions of each event or difficulty. Interviewer are trained to edit out the respondent's subjective reactions and emotional responses to the event, so that they cannot confound the event contextual rating. The event descriptions are then presented to a panel of raters, blind to the illness status of the respondent, who discuss the appropriate rating to assign the event or difficulty. Although in the earlier years of LEDS development, most events and difficulties were brought to these panels for discussion and rating, the availability of the detailed event and difficulty dictionaries ratings has routinized a significant proportion of the panel rating.

In addition to long-term contextual threat, a number of other dimensions are coded and rated for each event. These dimensions include type of event (e.g., illness, change in interaction); the person whom the event primarily affected or "focus" (including a "joint" category for events that happened to another but that are likely to have had an emotional impact on the respondent); the amount of threat reported by the respondent; and "independence," which is the probability that the event or difficulty could have been brought about by erratic behavior caused by the respondent's illness. The LEDS dictionaries also document the rules for determining these ratings.

The LEDS appears to be very reliable. One-year falloff in event reporting for *severe* events is about 1 percent per month (Brown & Harris, 1982), although the falloff for nonsevere and minor events is notable. Because the occurrence of a severe event (or presence of an ongoing severe difficulty) is found to be related to illness onset in their community studies (e.g., the estimated odds ratio ranges from 7 to 13 over a number of studies of depression; see Paykel, 1978; Surtees & Duffy, 1989), these investigators have argued that their method is more reliable than checklist measurement methods (e.g., Funch & Marshall, 1984).

The LEDS is an evolving instrument, with numerous studies now in progress to expand its reach to populations and illness onsets other than those that were the focus of its early development (schizophrenic patients and British working-class women). The LEDS has subsequently been used with adolescents (Monck & Dobbs, 1985) and the American elderly, with a new LEDS dictionary being developed for the latter (see note 1). A number of LEDS studies have been conducted in racial minority groups in England and Africa (for a review of the cross-cultural studies, see Brown & Harris, 1989b). Neither the LEDS nor LEDS-like methods, however, have been used with children. In addition to studies of schizophrenia and onsets of depression, the LEDS has been used in studies of anxiety, appendectomy, abdominal pain, menstrual disorders, multiple sclerosis, myocardial infarction, and dysphonia (for a review, see Brown & Harris, 1989a).

The LEDS has experienced criticism for its rating and interview methods. The most frequent and persistent criticism is that too much "context" is included in the ratings of contextual threat. Specifically, there is a long-standing controversy over whether its ratings of the magnitude of contextual threat muddle the distinction between the magnitude of event severity and the individual's "social" vulnerability to a stressor (Dohrenwend et al., 1993; Tennant, Bebbington & Hurry, 1981), thereby inflating estimates of event impact. An example makes this criticism clear. A death event can vary a great deal in severity in the LEDS rating scheme, from "below threshhold" for classification as an event to the highest category of severity. The severity rating depends, in part, on the reliance the person placed in the deceased as a confidant, and the number of alternative confidants available. Severity ratings for deaths also depend on other contextual factors that make a death more or less emotionally meaningful, such as closeness of kinship, quality of the predeath relationship, and frequency of contact the years before death (Brown, 1989). However, the inclusion of a social network vulnerability factor in the rating does confound concepts that some researchers may wish to keep separate (Tennant & Bebbington, 1978). The majority of event and difficulty ratings are less influenced by factors related to social vulnerability than is the death example; for instance, ratings of health events are based on objective contextual factors such as the life threat or disability posed by the illness, and its impact on work and other life routines (Brown, 1989). Similarly, ratings of work losses are based not on social vulnerability but rather on objective factors such as the unemployment rate, supply of jobs comparable to the one lost, and the proportion of family income lost.

Brown and his associates have continued to elaborate their measures of social vulnerability risk factors over time, and to a degree have met some of the criticism regarding the level of confounding between event severity and social vulnerability (Bebbington, 1986). This has been done through the development of separate interview measures of important social vulnerability factors such as commitment to role domains where events may arise, self-esteem, quality of social support, and conflict among social roles (Self-Evaluation and Social Support Schedule: Brown & Harris, 1989b). Nor, it should be noted, are *all* social vulnerabilities included in event ratings. Some aspects of social vulnerability, such as separation from a parent in childhood, are not included in event and difficulty ratings, but are measured and analyzed as additional risk factors for the onset of illness.

The LEDS has not come into general use, and has only recently been used in the United States. A number of European and American researchers have developed alternative instruments; some of them are structured-question elaborations of the life events checklist with contextual probes (e.g., Kessler & Wethington, 1991; Paykel, 1983; Wittchen, Essau, Hecht, Teder, & Pfister, 1989), whereas others are elaborations of the magnitude of "life change" rating systems (Dohrenwend et al., 1993). Other researchers have fielded shortened or modified versions of the LEDS (e.g., Costello & Devins, 1988; Miller et al., 1989; Wethington, Kessler, & Brown, 1993).

There are two explanations for why researchers have sought alternatives to the LEDS. The first is that many researchers are deterred by the cost and complexity of using the semi-structured LEDS interview. Important questions remain about the logistics of combining LEDS methods with the standardized methods of community epidemiological research, including the required 8 days of interviewer training in London or Pittsburgh (see note 1), the complicated rating scheme, and periodic retraining necessary in order to maintain consistent quality of interviews. Respondent burden is also an issue. Interviews are much lengthier than those required for checklist measures of life events (Katschnig, 1986). They also require a level of self-disclosure some respondents and interviewers find uncomfortable. The expense of the panel rating method, moreover, has been viewed as a significant deterrent.

The second reason is that some researchers prefer to make a stricter distinction between magnitude of change brought about by the event and vulnerability, an aim more compatible with the change–readjustment paradigm underlying checklist research methods (see Chapter 2). Recently, moreover, a few investigators have reported that the contextual rating system used in the LEDS may be no more predictive of illness onset that a carefully applied life-change unit rating scheme applied to events reported in a face-to-face interview (e.g., Faravelli & Ambonetti, 1983; Katschnig, 1980, 1986). In addition, within-respondent comparisons of events reported to life-change measures and to a LEDS interview suggest that the two methods yield different event reports (Katschnig, 1980; McQuaid et al., 1992), both of which could be related to onset of illness (Katschnig, 1980). At the time of this review, these questions remain to be more thoroughly investigated.

Alternative Interview and Rating Systems for Life Events

The Standardized Event Rating System

The Standardized Event Rating System (SEPRATE: Dohrenwend et al., 1993) is an alternative life-event interview and rating system, using a magnitude of "life change" rating system.[2] The SEPRATE, derived from the PERI life-events checklist (Dohrenwend, Krasnoff, Askenasy, & Dohrenwend, 1978) consists of a series of yes/no questions regarding 84 types of events or difficulties that may have occurred and been severely stressful. Each "yes" answer is probed with a number of structured questions, designed to fit a variety of circumstances. Although probes are provided, interviewers are not confined to the exact wordings as written. The purpose of the probes is twofold: to provide a narrative description of each event or difficulty, and to produce a standardized assessment of separate aspects of situations thought to produce experienced stress. These assessments include magnitude of change brought about by the event, desirability of the event (from the average, or "normative" perspective), disruptiveness to daily life routines, threat to life, independence from the respondent's own (possibly) disordered behavior, and "fatefulness" (occurring independently of the respondent's behavior and beyond his or her control to avert) (Dohrenwend et al., 1990).

After the interview is completed, an editor composes a narrative description of each event or difficulty reported. These descriptions differ from those done using the LEDS method, in that they are stripped of any material that might be used to infer "social vulnerability" to especially severe effects from the given stressor. This is done in order to prevent this sort of vulnerability from being inadvertently included in the rating of the event (Dohrenwend et al., 1990). Each event is then rated independently by two raters on the dimensions listed previously (Shrout et al., 1989). It is important to note that not all event context is stripped from the narrative. Indeed, many of the contextual factors taken account of in the LEDS are reproduced by the SEPRATE, such as the life-threatening or disabling aspects of an illness, the length of an unemployment episode, and the objective likelihood of being able to replace a lost job with one paying as well (Dohrenwend, 1993, personal communication).

Dohrenwend and associates have conducted a reliability study comparing the SEPRATE to their previously developed PERI normative life-event rating checklist (Shrout et al., 1989). Falloff for severe events is relatively modest (Raphael et al., 1991). The relative risk of onset of depression associated with "fateful" life events occurring in the previous 12 months is 1.50, and 3.38 for events determined to be both "fateful and highly disruptive" to life routines. The odds ratio associated with fateful and disruptive events is approximately double that estimated by use of the PERI checklist method in the same study (Shrout et al., 1989).

Both the rate of falloff and the case-control odds ratio estimated for onset appear somewhat unfavorable compared to the LEDS. However, it should be noted that the estimated odds-ratio provided by Dohrenwend and associates may have been driven down by including events and onsets reported to have occurred nearly 3 years before the interview, a longer than usual recall period. Indeed, a small case-control study (less than 25 subjects) fielded by Mazure and Bruce (M. Bruce, 1993, personal communication) indicates that the estimate of relative risk using the SEPRATE is comparable to the LEDS over a recall period more typical of studies of illness onset. It is also the case that rating dimensions are not as fine-tuned in the SEPRATE (see also below), perhaps leading to underrating of the potentially severe events. There is also the possibility that the SEPRATE is measuring a set of stressful experiences different from that which the LEDS typically evokes (see Katschnig, 1980). Theoretical refinement of event rating, combined with a shorter recall period, has produced odd-ratio estimates comparable to the LEDS (Dohrenwend, 1994, personal communication).

The SEPRATE is somewhat less comprehensive than the LEDS. The list of events and difficulties is less extensive in it as well as in other interview methods, reviewed below, that have been developed more directly from the LEDS. In particular, questions regarding important revelations about other people's character, bad news, decisions, and disappointments—an innovative feature in the LEDS—are not included. Questions such as these yield a number of severe events (Brown & Harris, 1978).

Like the LEDS, the SEPRATE technique has been successfully adapted for use in a different culture (Dohrenwend, 1993, personal communication). It has been used in Israel, where a different list of events and difficulties was utilized, although the rating dimensions remained the same. (These different events and difficulties reflected Israeli experience with war and terrorist activities.) The SEPRATE has been used primarily with adult community samples and with patients.

It is unclear whether the use SEPRATE interview may result in cost savings compared to the LEDS; however, not all information is available for comparison. In one study using the SEPRATE, interview length ranged from 60 to 90 minutes (Lennon, Dohrenwend, Zautra, & Marbach, 1990). Time estimates for narrative production, coding, and rating, however, are not available on a per interview basis. Training in the method is extensive, and (like the LEDS) effective use would require that investigators consult with the developers of the SEPRATE (Dohrenwend, 1993, personal communication).

Other Interview Measures

Other qualitative probe elaborations of event checklists have also been fielded by a number of investigators. These systems include the Detroit Couples Study Life Events Method (Kessler & Wethington, 1991); the Munich Events List (Wittchen et al., 1989); the Paykel Life Events Interview (Cooke, 1985; Paykel, 1983); and the Henderson, Byrne, and Duncan-Jones (1981) List of Recent Experiences. There are probably others that have escaped the detection of this review.

Fairly high reliabilities have been reported for some of these methods (although evaluations must be tempered because calculations of reliability are not consistent).

Falloff using these systems is also relatively modest for severe events. The use of these interviews, moreover, may result in significant cost savings.

The Detroit Couples Study Life Events Method. Like the SEPRATE, this interview is a checklist life event measure, with semi-structured probes administered after an event is reported. An offshoot of this interview was utilized in the National Survey of Health and Stress (Kessler, 1990).

Several features of the LEDS were included in its development. Acute, discrete events are differentiated from chronic difficulties. In addition to standard queries about domain-specific events, questions related to event "meaning" (e.g., revelations about the character of others, disappointments, situations turning out better than expected, and upsetting news) were adopted from the LEDS (Kessler & Wethington, 1991).

A number of features were incorporated into the design of the study in order to improve recall and accurate dating of life events. The features include concrete wordings of life event questions, contextual cues (arranging event questions by domain and embedding them in sections including other sorts of questions about that domain), multiple frames of reference (e.g., questions about learning something negative about the character of another person in addition to questions about difficulties "getting along" with someone), and memory aids (a life event calendar and calendar review to correct dates).

Four category stress severity ratings were assigned, based on the "loss" and "danger" long-term contextual threat ratings from the LEDS. However, the ratings were more similar to life-change event ratings than LEDS contextual threat ratings, since the interview did not collect sufficient contextual information to reproduce the latter.

A detailed analysis of event reports in the Detroit Couples study showed that this interview measure has acceptable psychometric properties overall, with some intriguing exceptions (Kessler & Wethington, 1991). The design of the study (husbands and wives interviewed from the same couple) made it possible to calculate reliability estimates for events rated as severe. The average reliability of reports was .64 for husbands and .68 for wives. However, inspection of reliability within event types showed that events that have the potential to be embarrassing, such as criminal activity and martial separations, were less reliably reported than other events. It is difficult to imagine that such a pattern reflects a common tendency to forget about embarrassing events. A more likely interpretation is that the occurrence of these events was consciously concealed. In addition, a comparison of husband and wife reports shows that *personal* events were consistently reported more accurately than events that occurred to one's spouse. The 2 to 3 percent monthly falloff rate for event reports, moreover, was lower than the 5 percent monthly falloff for a checklist reported by Funch and Marshall (1984), but greater than that reported for the LEDS (Brown & Harris, 1982).

The overall results, though, from the Detroit Couples Study are not as impressive as those obtained with the LEDS. The relative risk of onset of depression associated with severe events in the previous 12 months is 2.43 (odds ratio) compared to the range of 7 to 13 reported for the LEDS (see Paykel, 1978). It should be noted, moreover, that the estimate is also less than that obtained by the SEPRATE.

Nevertheless, the more structured methods of the interview may have resulted in a significant cost-saving. The average interview lasted 78 minutes. Coding time for the interview, including detailed coding of the life events, was less than 2 hours. Conventionally trained survey interviewers and coders were used, and only 1 day of special training was necessary for each group. Most significantly, 1755 interviews were conducted in a 4½-month period, and the coding of all events was accomplished in 2 months, a pace that could not even be approached using the LEDS.

It is important to note that methods such as these were developed specifically for a large-scale community survey. The measures used in the National Survey of Health and Stress (Kessler, 1990), an offshoot of the Detroit Couples Study, were applied in a sample of 8,000 representative of Americans, including ethnic minorities, aged 15 to 54.

The Munich Events List. The Munich Events List (Wittchen et al., 1989) is a three-step personal interview procedure. The interview consists of 85 life event descriptions, which are concretely specified. In the first step, the interviewer and respondent read the event descriptions together, with the respondent deciding whether such an event had occurred over the recall period. In the second step, the interviewer asks questions to assess the relevant context of the events already reported. In the third step, respondents rate the impact of each incident subjectively; then the interviewer rates the events on a number of objective dimensions, including whether the event involved gain or loss, was primarily positive or negative in its impact, could have been averted by the victim, or might have been brought on by the subject's mental condition or disordered behavior. There is also an objective severity rating scale, conceptually more similar to a life change–readjustment unit measure than the severity ratings used in the LEDS. Subjective and objective ratings of severity are correlated at .67 (Wittchen et al., 1989).

Reliability estimates were calculated in a test–retest analysis, by event type and/or severity. The investigators reported nearly perfect concordance for severe events such as deaths (kappa not reported), with lower agreement for less severe incidents (kappa = .74). Falloff was also estimated to be near zero for the severest events (0.36% per month, vs. 1% per month for the LEDS). Data have not been reported with which to calculate the 12-month risk of depression or other illness onset associated with severe events.

No data are available on the relative costs involved in fielding this interview, which has been used only in Europe. Samples have been small (Wittchen et al., 1989).

The Paykel Brief Life Event List. Paykel's life event interview (1983) has been used in a number of studies. This interview consists of 63 life event and difficulty questions, and can be administered in an average of 30 to 40 minutes (Cooke, 1985). Once an event or difficulty has been reported, the interviewer uses semi-structured probes in order to obtain more objective information and to estimate its emotional impact on the subject. Events and difficulties are dated, and it is determined whether they occurred before or after illness onset. They are also evaluated for their independence from the subject's illness status, whether the victim could have averted the event, respondent-reported negative and positive impact of the event, and objective contextual severity of impact, a measure adapted from the LEDS (Paykel, 1983). Interviewers are given 3 weeks of intensive training in rating events, and all events are rated by two independent raters. (It has not been reported how much the double-rating system reduces production time per interview.)

Inter-rater reliability of dimensions rated using this method appears to be adequate, ranging from .95 for specific event occurrence to .76 for objective negative impact (Paykel, 1983). In Cooke's replication (1985) inter-rater agreement on objective severity of impact (kappa = .64) is just within adequate range. The instrument has been used with adult community samples and patients.

Henderson, Byrne, and Duncan-Jones List of Recent Experiences. The LRE (Henderson et al., 1981; Steele, Henderson, & Duncan-Jones, 1980) is a personal interview method based primarily on the Tennant and Andrews (1976) life events scale. Probes administered after event responses are designed to date the event, distinguish discrete events from reports of chronic situations, obtain a brief description of the experience, and obtain ratings reflecting the respondent's judgment regarding severity of impact.

A separate staff rates events. Most of the ratings are based on the life-change weights developed by Tennant and Andrews (1976). However, for rare or unusual events for which no life-change weights are available, event weights are calculated by averaging the subjective impact reports of all sample members who reported that type of event. This averaging method is intended to reduce the confounding between subjective retrospective accounts of events and the respondent's mood.

Specific data about falloff in reporting severe events and question reliability are not available for a 12-month recall period, although 10-day test–retest reliability estimates are very good, estimated at .89 to .94 for the scale score, and .70 for specific events (Steele et al., 1980). The relative risk of *onset* of depression or other psychological disorder associated with severe events also cannot be calculated, since this study used a scale of "neurotic symptoms" as its outcome. However, the relative risk of experiencing a high level of neurotic symptoms among those reporting a high score on the life event scale (vs. a low level) was 2.88 (Henderson et al., 1981). The measure has been used only with adults.

Summary of Alternative Measures

Although the use of alternative interview measures of life events would substantially reduce the costs of conducting life event interviews, the present state of research on these instruments does not recommend their use without qualification. For example, the expectation that more structured life-event interviews would reduce interviewer and rating time is not necessarily fulfilled. The SEPRATE, which is one of the most highly developed of the alternative systems, requires extensive editing and expert rating upon its conclusion (Dohrenwend et al., 1993). Expert rating, as well as the preparation of event descriptions, requires a substantial investment of time that may be no less than that required to field the LEDS. The Paykel Brief Life Events List requires as much interviewer training as does the LEDS (Cooke, 1985); moreover, reliability estimates for the measure that maximally reduces interview costs are also disappointing (Kessler & Wethington, 1991).

Some alternative methods, however, do have their strengths. The Detroit Couples Study method, which does not require extensive interviewer training, is clearly the most compatible with existing community survey methods in the United States. The SEPRATE has several useful rating dimensions and is compatible with a widely accepted theory of the stress process. Falloff estimates for the Paykel Brief Life Events List and the Munich Life Events Inventory rival those reported for the LEDS. Unfortunately, there is no research yet published that directly compares these methods and weighs their theoretical and logistical costs and benefits.

Shortened Versions of the LEDS

Miller and Salter (1984) are perhaps best known for their conclusion that there is no "shortcut" for rating the severity of life events with the precision of the LEDS. Yet they also concluded that better life-event rating could be achieved through the use of more detailed standardized probing, rather than through continued reliance on unstructured probing.

In response to this statement, a number of investigators have attempted to improve the LEDS interview method by modifying its interview techniques, primarily by structuring probes for event, which would both shorten the time it takes to administer the interview and determine the event ratings. Such structuring, if demonstrated to reproduce the event ratings produced by the LEDS, would substantially reduce its costs. Three revisions of the LEDS are reviewed in the following sections.

The Structured Life Events Inventory

The Structured Life Events Inventory (SLI: Wethington, Kessler, & Brown, 1993) was designed as a *more structured* version of the open-ended LEDS.[3] The purpose of creating the structured version was to experiment with ways to reproduce Brown and Harris's contextual threat ratings, while also using interviewing techniques more consistent with conventional American survey techniques (e.g., the Detroit Couples Study). Another purpose of structuring the LEDS was to reduce the amount of time necessary in order to produce contextual threat ratings, as well as to cut the amount of necessary training time.

The SLI resembles traditional survey interviews more than it does the LEDS, but it still differs significantly from them as well. Consisting of a series of questions designed to elicit events, the SLI provides instructions for the interviewers to probe some of the responses in an extemporaneous, conversational manner, that is, to follow the lead of probe questions that are suggested in the interview, but not necessarily to read the questions verbatim.

As does the LEDS, the SLI requires interviewers to make research judgments while they are interviewing. Interviewers are told that the purpose of the probing is to estimate the objective severity of an event or difficulty reported by the respondent. During the interview, interviewers make judgments about the severity of the event or difficulty the respondent is reporting, and record their summary judgment regarding severity in the interview, along with a detailed description of the situation. Because both of these tasks require fairly precise knowledge about the event or difficulty, as well as some understanding of some of the important aspects of the respondent's life that may increase the event's severity, care has been taken to assist the interviewers in making correct judgments. Time-consuming probing of events and difficulties that are of low-moderate or less severity, as well as failing to detect potentially severe events and difficulties, are errors that can greatly reduce the quality of the data.

Although this set of interviewer tasks and judgments may initially seem very challenging, and very difficult to train and supervise, the investigators tried to reduce the burden on the interviewers in several ways—primarily by reducing reliance on memory. A training manual was produced that described the general principles used to make judgments about event and difficulty severity. The conventionally trained survey interviewers were given 3 days of training. Exercises based on examples reported in the LEDS dictionaries were used to teach the interviewers how to make judgments. Interviewers were also trained in probing techniques.

In order that training time be reduced, the interview schedule contained as many instructions as feasible to make it possible for determining severity. This was done so that interviewers would not have to memorize a great deal of specific information about what sorts of events and situations are rated as severely stressful, or less. In the interview design, two major strategies were utilized. The first principle was to begin probe sequences with questions designed to screen out as quickly as possible events that were not likely to be severely stressful. For example, reported financial difficulties were not probed if respondents had not had to deplete their savings, sell or skip payments on their major possessions, or go without medical care and necessities such as food and clothing for the family. If events were determined not to reach threshold for severity, the interviewer was instructed to skip to the next question sequence.

The second design principle was to produce for each question sequence a series of optional probes that were sensitive to contextual variations that make an event severe or not. An example of this principle is the question and probe sequence for respondent pregnancy: Did this pregnancy have any negative impact on your career or job plans for the future? (Optional: (Will you be/Were you) able to keep your job? (Will you be/Were you) forced to cut back on your work hours? [IF YES] (Is/Was) this your choice or someone else's decision? (Did you have/Will you have) to turn down a promotion, transfer to a less demanding job, or put your career plans "on hold" in any way? [IF YES] (Is/Was this your choice, or someone else's decision?) In addition to the sequence excerpted, there are also questions about whether there were difficulties with the father or other family member during the pregnancy, financial or housing problems, near miscarriages, or other medical emergencies involving either mother or baby. Each includes a set of optional probes to gather information about potentially severely stressful circumstances surrounding the event.

In order to examine interviewer judgment and maintain consistency, all interviews are then examined and rerated by an expert rater, trained in the LEDS method. Ratings for long-term contextual threat, event type, independence from respondent illness, and focus are assigned to the events, based on interviewer ratings and other information from the interview.

The SLI was fielded in an experimental study in 1992. This study was designed to examine the reliability and validity of the structured version, in comparison with the original semi-structured method. A sample of 243 community-dwelling residents were interviewed, half with the SLI and half with the original LEDS. Forty-one of the respondents were interviewed twice, using both methods. The LEDS interviewing and weekly panel consensus meetings were supervised by an expert investigator trained by Brown and Harris.

Results from this comparison study, although still under review, are promising. The average SLI interview production time (including interviewing, coding, and rating) was 9 hours, compared to 16 hours required for the LEDS. SLI interviewers can conduct more interviews per week (8) than LEDS interviewers (3–4). In addition, SLI interviewers were able to reliably distinguish events of severe long-term contextual threat from more minor events. In addition, the risk of depression onset associated with severe events and difficulties (3 months prior to onset) was similar for the LEDS and SLI in this study, with the LEDS having a slight advantage.

Nevertheless, much work still needs to be done on the SLI. Between-interviewer rating reliability was not perfect. The LEDS appears to be superior to the current SLI in eliciting and rating difficulties. In order to improve the SLI, development of a second experimental test is underway.

Costello's Two-Stage Screening Method

Costello (1982) and Costello and Devins (1988) have experimented with a two-stage screening method in order to reduce LEDS interviewing and rating time. Reasoning that some of the discrepancies found between self-administered checklist and personal interview techniques were due to discrepancies in how *events* were defined, the investigators developed a checklist based on the LEDS. Event and difficulty descriptions were worded to reflect contextual circumstances that increase severity ratings.

Study participants were asked to fill out a screening checklist. Within a week of completing the screening, participants were interviewed in their homes by LEDS-trained interviewers who were blind to the checklist results. The investigators then compared severe events and difficulties reported from the two methods.

The interview and checklist methods were highly consistent when a severe event or difficulty had *not* occurred. However, the two techniques were not as consistent when such an event had occurred; only 27 percent of those indicating a "severe" event or difficulty on the checklist were rated as having experienced a severe event or difficulty of that type in the subsequent interview. These results indicate, though, that a checklist method could be used to screen for events at the beginning of a personal interview, and that the subsequent LEDS administration could be short-ened. Such shortening would save considerable interviewing and rating time.

The Edinburgh Version of the LEDS

Miller and associates (Miller et al., 1986) have also fielded a shortened version of the LEDS. As recommended by Costello (1982), this interview uses a self-administered checklist screening procedure, with the interviewer probing only those events checked on the list. A panel then assigns ratings based on the LEDS system.

In a study using this method, the relative risk of onset of psychiatric disorder (depression or anxiety) is estimated at 2.40 for those who experienced any severe event and 3.21 for those who experienced a severe event or difficulty (Surtees et al., 1986). Falloff rates for event recall appear to be very similar to the LEDS (Surtees et al., 1986). This methods appears to have shortened interview and rating time, although per-interview time estimates have not been published.

Future Directions

It is traditional to end such a review with a call for further research. There is a need for more research comparing the interview methods themselves, and more attention to documenting the costs of different personal interview methods. More critical, though, are guidelines to help researchers make judgments about *which* of the methods reviewed is appropriate for their projects.

Choosing the Appropriate Method

Two critical criteria are necessary for judging the appropriateness of an interview method: (1) its suitability to answer the primary research questions of the study, and (2) practical limitations.

Several unresolved issues that have circumscribed use of personal interviews based on the LEDS still remain. The first is that the rating system, although clearly defined and documented, confounds some event severity ratings with social vul-nerability factors. Many have argued (e.g., Dohrenwend et al., 1993) that these dimensions should be kept strictly separate. The second issue is that the costs of administering any personal interview measure are high in comparison to a checklist. A number of researchers have therefore sought alternatives in order to keep costs down. The third issue is interviewing logistics, over and above study costs. Studies utilizing the LEDS have employed relatively small samples (usually 400 or less). Interviewers require extensive training. Post-interview rating and attendance at weekly panel meetings limit the number of interviews that can be completed per week; as a result, interviews take place only over long periods of time, or a very large number of interviewers must be hired and trained. Even if the panel method is not used to rate events and difficulties, as in the SLI (Wethington et al., 1993) and the Brief Life Event List (Paykel, 1983), interviewers must be closely supervised and periodically retrained. Some progress has been made in addressing the confounding issue of rating, primarily through the development of the SEPRATE. It is still unclear, though, whether the SEPRATE represents a significant cost savings over the LEDS, or reduces the logistical problems that make the LEDS less practical to apply in larger sample surveys. The ready availability of the LEDS dictionaries, the longer time in which the LEDS has been used, and its ready application to diverse populations (e.g., adolescents and other cultural groups)—all these advantages may outweigh the confounding issue of rating in some researchers' eyes.

Significant progress has also been made in reducing cost and logistical problems, primarily through the development of shortened or more standardized versions of the LEDS (Cooke, 1985). Questions still remain, however, about their replicability and reliability, since none has been used as extensively as the LEDS.

Acknowledgments

The development of this chapter has been supported by grants from the MacArthur Foundation Research Network on Successful Midlife Development to Ronald C. Kessler and USDA Hatch (321-7419) to Elaine Wethington. The authors wish to thank Sheldon Cohen, Stanislav Kasl, Bruce Dohrenwend, Bruce Link, and Martha Bruce for their comments and contributions to this chapter.

Notes

1. Training for the LEDS is conducted by Tirril O. Harris at Royal Holloway and Bedford New College, 11 Bedford Square WC1, UK; and also by Dr. Ellen Frank, University of Pittsburgh, Western Psychiatric Institute and Clinic, Bellefield Tower, 3811 O'Hara Street, 8th floor, Pittsburgh, PA, 15213, USA.

2. Training for the SEPRATE is available through a staff supervised by Dr. Bruce Dohrenwend, New York State Psychiatric Institute, Box 8, 722 West 168th Street, New York, NY 10032, USA.

3. Inquires about use of the SLI should be directed to Dr. Elaine Wethington, Department of Human Development and Family Studies, G63 Van Rensselaer Hall, Cornell University, Ithaca, NY 14853, USA.

References

Bebbington, P. (1986). Establishing causal links: Recent controversies. In H. Katschnig (Ed.), *Life events and psychiatric disorder: Controversial issues* (pp. 188–200). Cambridge: Cambridge University Press.

Bebbington, P. E., Brugha, T., MacCarthy, B., Potter, J., Sturt, E., Wykes, T., Katz, R., & McGuffin, P. (1988). The Camberwell Collaborative Depression Study. I. Depressed probands: Adversity and form of depression. *British Journal of Psychiatry, 152,* 754–765.

Brown, G. W. (1981). Contextual measures of life events. In B. S. Dohrenwend & B. P. Dohrenwend (Eds.), *Stressful life events and their contexts* (pp. 187–201). New York: Prodist.

Brown, G. W. (1989). Life events and measurement. In G. W. Brown & T. O. Harris (Eds.), *Life events and illness* (pp. 3–45). New York: Guilford.

Brown, G. W., Adler, Z., & Bifulco, A. (1988). Life events, difficulties, and recovery from chronic depression. *British Journal of Psychiatry, 152,*487–498.

Brown, G. W., Bifulco, A., & Harris, T. O. (1987). Life events, vulnerability, and onset of depression: Some refinements. *British Journal of Psychiatry, 150,* 30–42.

Brown, G. W., & Harris, T. O. (1978). *Social origins of depression: A study of depressive disorder in women.* New York: Free Press.

Brown, G. W., & Harris, T. O. (1982). Fall-off in the reporting of life events. *Social Psychiatry, 17,* 23–28.

Brown, G. W., & Harris, T. O. (1989a). *Life events and illness.* New York: Guilford.

Brown, G. W., & Harris, T. O. (1989b). Summary and conclusions. In G. W. Brown & T. O. Harris (Eds.), *Life events and illness* (pp. 439–482). New York: Guilford.

Cannell, C. F., Miller, P. V., & Oksenberg, L. (1981). Research on interviewing techniques. In S. Leinhardt (Ed.), *Sociological methodology* (pp. 389–436). San Francisco: Jossey-Bass.

Cooke, D. J. (1985). The reliability of a brief life event interview. *Journal of Psychosomatic Research, 29,* 361–365.

Costello, C. G. (1982). Social factors associated with depression: A retrospective community study. *Psychological Medicine, 12,* 329–339.

Costello, C. G., & Devins, G. M. (1988). Two-stage screening for stressful life events and chronic difficulties. *Canadian Journal of Behavioral Science, 20,* 85–92.

Dohrenwend, B. S., Krashnoff, L., Askenasy, A. R., & Dohrenwend, B. P. (1978). Exemplification of a method for scaling life events: The PERI life events scale. *Journal of Health and Social Behavior, 19,* 205–229.

Dohrenwend, B. P., Link, B. G., Kern, R., Shrout, P. E., & Markowitz, J. (1990). Measuring life events: The problem of variability within event categories. *Stress Medicine, 6,* 179–187.

Dohrenwend, B. P., Raphael, K. G., Schwartz, S., Stueve, A., & Skodol, A. (1993). The structured event probe and narrative rating method for measuring stressful life events. In L. Goldberger & S. Breznitz (Eds.), *Handbook of stress: Theoretical and clinical aspects* (pp. 174–199). New York: Free Press.

Faravelli, C., & Ambonetti, A. (1983). Assessment of life events in depressive disorders: A comparison of three methods. *Social Psychiatry, 18,* 51–56.

Fiedler, K., & Stroehm, W. (1986). What kind of mood influences what kind of memory?: The role of arousal and information structure. *Memory and Cognition, 14,* 181–188.

Funch, D. P., & Marshall, J. R. (1984). Measuring life stress: Factors affecting fall-off in the reporting of life events. *Journal of Health and Social Behavior, 15,* 453–464.

Gorman, D. M., & Brown, G. W. (1992). Recent developments in life-event research and their relevance for the study of addictions. *British Journal of Addiction, 87,* 837–849.

Harris, T. O. (1991). Life stress and illness: The question of specificity. *Annals of Behavioral Medicine, 13,* 211–219.

Henderson, S., Byrne, D. G., & Duncan-Jones, P. (1981). *Neurosis and the social environment.* New York: Academic Press.

Katschnig, H. (1980). Measuring life stress: A comparison of two methods. In R. Farmer & S. Hirsch (Eds.), *The suicide syndrome* (pp. 116–123). London: Croom Helm.

Katschnig, H. (1986). Measuring life stress: A comparison of the checklist and panel technique. In H. Katschnig (Ed.), *Life events and psychiatric disorders: Controversial issues* (pp. 74–106). Cambridge: Cambridge University Press.

Kessler, R. C. (1990). *The National Survey of Health and Stress.* Ann Arbor, MI: Survey Research Center, University of Michigan.

Kessler, R. C., & Wethington, E. (1991). The reliability of life event reports in a community survey. *Psychological Medicine, 21,* 723–738.

Lennon, M. C., Dohrenwend, B. P., Zautra, A. Z., & Marbach, J. J. (1990). Coping and adaptation to facial pain in contrast to other stressful events. *Journal of Personality and Social Psychology, 59,* 1040–1050.

McQuaid, J., Monroe, S. M., Roberts, J. R., Johnson, S. L., Garamoni, G. L., Kupfer, D. J., & Frank, E. (1992). Toward the standardization of life stress assessment: Definitional discrepancies and inconsistencies in methods. *Stress Medicine, 8,* 47–56.

Miller, P. McC., Dean, C., Ingham, J. G., Kreitman, N. B., Sashidharan, S., & Surtees, P. G. (1986). Life events and long term difficulties in a sample of Edinburgh women, with some reflections on the concept of independence. *British Journal of Psychiatry, 148,* 686–696.

Miller, P. McC., & Salter, D. P. (1984). Is there a short-cut? An investigation into the life event interview. *Acta Psychiatrica Scandinavica, 70,* 417–427.

Monck, E., & Dobbs, R. (1985). Measuring life events in an adolescent population: Methodological issues and related findings. *Psychological Medicine, 15,* 841–850.

Paykel, E. S. (1978). Contribution of life events to the causation of psychiatric illness. *Psychological Medicine, 8,* 245–253.

Paykel, E. S. (1983). Methodological aspects of life event research. *Journal of Psychosomatic Research, 27,* 341–352.

Raphael, K. G., Cloitre, M., & Dohrenwend, B. P. (1991). Problems with recall and misclassification with checklist methods of measuring stressful life events. *Health Psychology, 10,* 62–74.

Shrout, P. E., Link, B. G., Dohrenwend, B. P., Skodol, A. E., Stueve, A., & Mirotznik, J. (1989). Characterizing life events as risk factors for depression: The role of fateful loss events. *Journal of Abnormal Psychology, 98,* 460–467.

Sobell, L. G., Toneatto, T., Sobell, M. B., Schuller, R., & Maxwell, M. (1990). A procedure for reducing error in reports of life events. *Journal of Psychosomatic Research, 2,* 163–170.

Steele, G. P., Henderson, S., & Duncan-Jones, P. (1980). The reliability of reporting adverse experiences. *Psychological Medicine, 10,* 301–306.

Surtees, P. G., & Duffy, J. C. (1989). Binary and rate measures of life event experience: Their association with illness onset in Edinburgh and London community surveys. *Journal of Affective Disorders, 16,* 139–149.

Surtees, P. G., Miller, P. McC., Ingham, J. G., Kreitman, N. B., Rennie, D., & Sashidharan, S. P. (1986). Life events and the onset of affective disorder: A longitudinal general population study. *Journal of Affective Disorders, 10,* 37–50.

Tennant, C., & Andrews, G. (1976). A scale to measure the stress of life events. *Australian and New Zealand Journal of Psychiatry 10,* 27–32.

Tennant, C. & Bebbington, P. (1978). The social causation of depression: Critique of the work of Brown and his colleagues. *Psychological Medicine, 8,* 565–575.

Tennant, C., Bebbington, P., & Hurry, J. (1981). The role of life events in depressive illness: Is there a substantial causal relation? *Psychological Medicine, 11,* 379–389.

Wethington, E., Kessler, R. C., & Brown, G. W. (1993). *Training manual and technical report for the structured life events inventory.* Ithaca, NY: Life Course Institute, Cornell University.

Wittchen, H., Essau, C. A., Hecht, H., Teder, W. & Pfister, H. (1989). Reliability of life event assessments: Test–retest reliability and fall-off effects of the Munich Interview for the Assessment of Life Events and Conditions. *Journal of Affective Disorders, 16,* 77–91.

Zimmerman, M. (1983). Methodological issues in the assessment of life events: A review of issues and research. *Clinical Psychology Review, 3,* 339–370.

4

Daily and Within-Day Event Measurement

John Eckenrode and Niall Bolger

In this chapter we review methods that assess stressful experiences at the daily level, primarily through self-reports using some form of daily diary or record. Although other forms of self-recording devices are available to measure bodily states, such as blood pressure (see Chapter 10), these techniques are less relevant to the assessment of exposure to stressors or the cognitive or behavioral responses to stress and, as such, will not be emphasized here.

There are compelling reasons why a researcher may choose to assess stressful experiences and related constructs such as coping at the daily level (Stone & Shiffman, 1992; Verbrugge, 1980). Measurement at the daily level affords the researcher with the equivalent of a behavioral science microscope. By recording the details of human thoughts, feelings, and actions, diaries have a resolving power that cannot easily be achieved through the use of standard interviews or questionnaires. For stress researchers, this resolving power can be used to tackle important theoretical problems such as the nature of chronic stress, the mechanisms through which major stressors exert their effects, and the role of personality and social structure in the stress process.

Traditionally, to measure chronic and acute stressors, stress researchers obtained retrospective reports stretching back over months and years. Such methods are prone to a variety of recall biases that limit their accuracy. Although they are not applicable to all types of stress research, diary designs can be used prospectively to study chronic and acute stressors over periods of weeks and months. Furthermore, as theories of stress focus increasingly on intervening processes and mechanisms (Kessler, Price, & Wortman, 1985), the value of methods that can tap into everyday thoughts, feelings, and actions increases. Finally, the prospective nature of diary data helps to address some of the causal ambiguity inherent in retrospective survey data.

Short History of Daily Life Event Research

Wheeler and Reis (1991) have recently reviewed the history of methods of self-recording of everyday life events. Additional historical references for the use of health diaries are provided by Verbrugge (1980). Excellent summaries of available methodologies for the daily assessment of stressful events have been published by Stone, Kessler, & Haythornthwaite (1991) and Stone and Shiffman (1992). Suls and Martin (1993) also provide a brief history of daily recording methods. What is clear from these reviews is that daily methods, while not playing a prominent role in behavioral science research, have enjoyed a rather long history and are becoming increasingly popular. The latter is evidenced by the appearance of two recent special issues of journals devoted to research using daily reports and self-recording methods (*Journal of Personality* 59:3, 1991; *Annals of Behavioral Medicine* 15:1, 1993).

Some of the early uses of daily records or diaries were in consumer expenditure surveys, studies of food consumption, and time use surveys (e.g., Sudman & Ferber, 1971; Szalai, 1972). Verbrugge (1980) reports even earlier uses of diaries in health studies, beginning with the Baltimore Morbidity Survey from 1938 to 1943 (Downes & Collins, 1940). In such early health studies, the diary was used primarily as a memory aid when researchers conducted interviews about illness episodes. More recent health studies have also used diaries, including the National Medical Care Expenditure Survey, in which respondents keep a diary of medical expenses. As Verbrugge (1980) points out, these health studies have included diaries primarily as a methodological check on the accuracy of data on health-related episodes and behaviors collected with more conventional interviews, questionnaires, or health records. Compared to these other methods, diaries generally yield higher incidence rates for acute and chronic conditions (Verbrugge, 1980).

In the field of psychology, Wheeler and Reis (1991) point to several trends that fueled an interest in methods to assess everyday experience. The renewed interest in inner experience in the 1960s along with the appearance of research instruments designed to assess momentary psychological states (e.g., the Nowlis Mood Adjective Checklist; Nowlis, 1965) have led to many studies that have explored the stability of mood over several days, the correlates of changes in mood, and the relation of mood changes to more major changes in physical or psychological health status (see Chapter 7).

Research and therapy based on behaviorist theories has also generated an interest in assessing daily experience. Research on psychological disorders, such as depression has benefited from examining the behavioral correlates of positive and negative mood states on a daily basis (e.g., Lewinsohn & Libet, 1972; Lewinsohn & Talkington, 1979). In the therapeutic context, diaries have served as a method for identifying behaviors that may be the focus of treatment and for evaluating the effectiveness of treatment strategies. In this tradition, the assumption was that the act of recording everyday thoughts, feelings, and behaviors may itself modify behavior (see Duncan, 1969). The reactive nature of these techniques, although viewed as a strength among the behavioral therapists, remains a source of concern among researchers using these methods primarily as a data collection tool.

Personality psychologists have also recognized the value of assessing daily experiences as a way to explore stability and change in personality constructs over time (e.g., Epstein, 1979, 1980). Csikszentmihalyi and his colleagues (e.g., Csikszentmihalyi, Larson, & Prescott, 1977) have utilized the "experience sampling method" (ESM) to tap into subjects' (usually adolescents') thoughts or behaviors at random intervals over several days. An electronic paging device is used as the signal for the subject to record (see Hormuth, 1986, for a review of these methods). These methods have been valuable in documenting the prevalence and stability of psychological states (e.g., negative affect; Larson & Ham, 1993) as well as behaviors (e.g., time studying) and in exploring the co-occurrence of these states and behaviors (e.g., Wong & Csikszentmihalyi, 1991).

One of the earliest uses of the diary method in stress research was a study by Meyer and Haggerty (1962) in which families kept diaries of stressful family events, that were then related temporally to the outbreak of streptococcal infections. This was followed by a community-based study by Roghmann and Haggerty (1972, 1973) in which a large sample of adults kept daily diaries of stressful events, illnesses, and the use of health services. These studies established some of the first links in the literature between minor daily stressors and health-related outcomes. Studies in the early 1980s by Eckenrode and colleagues (Caspi, Bolger, & Eckenrode, 1987; Eckenrode, 1984; Gortmaker, Eckenrode, & Gore, 1982); and by Verbrugge as part of the Health in Detroit Study (e.g., Verbrugge, 1985), used a simple, open-ended question or an abbreviated checklist to record stressful events, and then related these daily measures to mood, illnesses, or health behaviors.

At about this time, several other research groups began exploring daily stressors and their relation to psychological or physical health outcomes. Mirroring the earlier development and widespread usage of the Schedule of Recent Experiences (SRE: Holmes & Rahe, 1967), the measures of daily stressors typically involved checklists. Lewinsohn and associates developed the Pleasant Events Schedule to explore the relationship between depression and levels of daily positive experiences and activities (Lewinsohn & Libet, 1972), as well as its counterpart, the Unpleasant Events Schedule (Lewinsohn & Talkington, 1979).

Emerging from their transactional view of the stress process, Lazarus and his colleagues began to explore the nature of everyday stressors as a complement to major life events. This research group produced the 117-item Hassles Scale and the 135-item Uplifts Scale which originally were used as monthly measures (Kanner, Coyne, Schaefer, & Lazarus, 1981) but were also adapted for use as a daily measure (DeLongis, Folkman, & Lazarus, 1988). Although there has been some controversy concerning the extent to which the Hassles Scale is confounded with mental health outcomes (see Dohrenwend, Dohrenwend, Dodson, & Shrout, 1984; Lazarus, De-Longis, Folkman, & Gruen, 1985), its popularity has served to focus the attention of stress researchers on small life events. It has also served as the basis for the development of other measures that assess hassles generally or in specific domains, such as the Parenting Daily Hassles Scale (Crnic & Greenberg, 1990).

The early 1980s also saw the development of checklist measures specifically designed to be used on a daily basis. Stone and associates developed the Assessment of Daily Experience (ADE) measure (Stone & Neale, 1982). This 66-item scale included both positive and negative experiences distributed across five major domains (e.g., work-related activities, leisure activities). The current version, the Daily Life Experience (DLE) checklist, contains 78 items (Stone, Neale, & Shiffman, 1993). In addition to engaging in extensive piloting studies to generate scale items, these researchers have also reported concordance data between husband and wife couples as a way to address validity issues. Such checklist measures are reviewed in more detail below.

Currently, researchers are adapting these earlier efforts to address a range of theoretical issues in stress research. As more researchers have utilized daily approaches to the assessment of stressful experiences, discussions of methodological and analytic issues in daily studies have also become more commonplace (e.g., Jaccard & Wan, 1993; Kenny, Kashy, & Bolger, 1993; Stone, Kessler, & Haythornthwaite, 1991; West & Hepworth, 1991).

Research Questions That Can Be Addressed with Daily Event Measures

Daily event measures can be used to address a wide range of research questions. We list some of the major ones below and give examples of studies in each case. Note that we assume that the measures are used in a daily diary design—namely, a longitudinal design that involves repeated assessments of daily stressors over the course of a single day, over multiple days, or both. The daily data that are collected in such designs can be aggregated over days to obtain person-level stress measures, such as the total number of days on which work conflicts occurred. Such stress measures may be less prone to recall biases than retrospective reports covering that same time period, but it is questionable whether the costs of using diary data in this way outweigh any benefits in reducing measurement error. A much better use of diary data is to preserve the longitudinal nature of the data by keeping the analysis at the level of the person-day. Analyses at the daily level can examine temporal relationships between daily stressors and daily outcomes, while still allowing the researcher to examine individual differences in these processes. We discuss data analysis issues later in this chapter.

The first and most obvious use of daily event measures is to document people's exposure to stressors. In studies of work stress, for example, researchers may be interested in knowing how often workers are exposed to various types of daily problems such as work overloads and interpersonal conflicts with co-workers, supervisors, or subordinates. A recent study by Bolger et al. (1989) found that work overloads and work conflicts differed dramatically in their prevalence. Workers complained of being overloaded with work on 1 out of every 3 days, whereas they complained of having work conflict only once a month.

It is commonly believed that negative events can have a cascading effect such that "one bad thing leads to another." A second use of daily event measures is to investigate this belief by studying the interrelationships between daily events over time. For example, one way in which the management of work and family roles creates distress is that stressful events in one domain lead to stressful events in the other domain (Bolger, DeLongis, Kessler, & Wethington, 1989).

A third use of these measures, and the one that has received the most research attention, is to study the impact of daily events on physical and mental health. Initial work demonstrated associations between summary measures of daily events and psychological distress (Kanner et al., 1981). This work has been expanded to examine the emotional impact of specific categories of daily events (Bolger, De-Longis, Kessler, & Schilling, 1989; Stone, 1987). For example, the Bolger et al. (1989b) study mentioned earlier found that work conflicts are the most distressing of all daily events, whereas the much more common work overloads were not very distressing. Progress has also been made on understanding the relationship of daily events to physical symptoms (DeLongis, Coyne, Dakof, Folkman, & Lazarus, 1982; DeLongis et al., 1988).

A fourth type of question that can be addressed using daily event measures is the extent to which daily events make up the elements of chronically stressful experiences (see Chapter 5). For example, daily event measures can permit investigators to identify the discrete events (overloads, conflicts) that make up a chronically stressful marriage or occupation. Consistent with this idea, Eckenrode (1984) has shown that daily events are predicted by measures of chronic stress.

Daily events are also predicted by measures of major life events. This is thought to occur because major life events affect health outcomes, in part through increasing daily problems (Eckenrode, 1984; Kessler, Price, & Wortman, 1985). Therefore, a fifth use of daily event measures is to assess the extent to which they mediate the effects of major life events. For example, in a study of low-income mothers, Eckenrode (1984) showed that major life events predicted subsequent daily events which, in turn, predicted daily distress. Pearlin and colleagues (1981) demonstrated that the effects of major work disruptions such as unemployment include chronic role strains and thereby increased depression. As noted above, daily events are likely to be the key elements of chronic stress.

More generally, a fruitful area of research is the extent to which daily events mediate the effects of personality and social variables on disease. For example, daily events have been shown to explain partially the effects of personality variables, such as neuroticism, on daily distress (Bolger & Schilling, 1991) and physical symptoms (Larsen & Kasimatis, 1991).

Finally, as is the case of major events, people differ in the extent to which they are emotionally and physically affected by daily stressors. Researchers have begun to explore personality and social variables that modify the impact of these stressors. For example, the effects of daily stressors have been shown to depend on self-esteem (Campbell, Chew, & Strachey, 1991; DeLongis et al., 1988), neuroticism (Bolger & Schilling, 1991; Larsen & Kasimatis, 1991), and social support (Caspi, Bolger, & Eckenrode, 1987; DeLongis, Folkman, & Lazarus, 1988).

All of the above questions can be answered within a data-analytic framework variously known as the multi-level model, the mixed model, the hierarchical linear model, and the covariance components model. The essential feature of these models is that they recognize that diary data have two random components: one due to the sampling of persons (or couples, families, etc.), and the other due to the sampling of repeated measurements within persons (e.g., days). To make correct inferences from diary data, both of these sources of variance need to be considered. Using statistical models that fail to take both random components into account (e.g., conventional regression and ANOVA models) can produce very misleading results. Kenny, Kashy, and Bolger (1993) provide an overview of analysis approaches to diary data and explain in detail why it is necessary to use multi-level analytic models. Guidance on diary data analysis is also provided by West and Hepworth (1991) and Jaccard and Wan (1993).

Software for correctly analyzing these data is available mostly in specialized programs such as HLM (Bryk & Raudenbush, 1992), ML3 (Prosser, Rasbash, & Goldstein, 1991), GENMOD (Mason, Anderson, & Hayat, 1988). MIXREG (Hedeker, 1993), and VARCL (Longford, 1986). Equivalent programs are also available within standard statistical packages: MIXED is available as part of SAS software, and 5V is available as part of BMDP software. All these programs use a multiple-step Maximum Likelihood approach to estimation and testing. Kenny, Kashy, and Bolger (1993) describe a simpler, one-step Weighted Least Squares approach that can be carried out using SAS's GLM program.

Choosing an Appropriate Measure

It is useful to categorize methods used in daily stress assessment as an aid to researchers who are considering using a daily approach. Such a taxonomy will also be useful in pointing out the limitations of this review, which is focused primarily on a subset of all possible methods.

Wheeler and Reis (1991) describe three generic types of daily self-report methods: (1) interval-contingent recording; (2) signal-contingent recording; and (3) event-contingent recording. Suls and Martin (1993) added a fourth category to this scheme, continuous recording, which will not be discussed here since it is less relevant to stress research. Although interval-contingent recording has been by far the most frequently used method in stress research, we describe both signal-contingent and event-contingent methods because of their potential benefit to researchers in this field.

Interval-Contingent Recording

In this method, data are collected at regular intervals, determined ahead of time by the researcher. Typically the time interval has been a day, although shorter and longer time periods have been used. The respondent will usually report on events that have occurred since the last self-report, although specific research questions may require that events covering a shorter period of time between recordings be assessed. For example, in her study of work stress and marital interaction among air traffic controllers and their spouses, Repetti (1989) had respondents complete daily diaries that asked them to report on marital behavior for that evening, since the research question concerned the effects of the work day on subsequent marital interactions.

Signal-Contingent Recording

This method involves having respondents record their experiences whenever signaled by the researcher. Typically, the interest is with what the respondent is experiencing at the precise moment they are signaled. As such, the time period covered is very brief. A good example of this method is the Experience Sampling Method (ESM) developed by Csikszentmihalyi and Larson (Csikszentmihalyi & Larson, 1987; Larson & Csikszentmihalyi, 1983). In their work, electronic "beepers" worn by respondents are used to signal the recording period. Respondents may be signaled several times a day, at random or during prespecified times. These methods may be particularly useful in recording inner experiences, such as mood, that may fluctuate often during the day, and thus be subject to recall biases if assessed several hours or days later.

Event-Contingent Recording

This method requires respondents to make a report every time a predetermined event has occurred. For example, in research with the Rochester Interaction Record (Wheeler & Nezlek, 1977) and the Iowa Communication Record (Duck, 1991), respondents are instructed to use diaries or logs to answer questions about social interactions as they occur. Depending on the research question being addressed, the event may be a predictor variable, such as an argument with a spouse, or an outcome variable, such as a headache or other physical symptom. Since the events in question may occur at any time, the length of time between records varies. Such an approach is well suited to events with a well-defined beginning and ending, such as an argument, but is less appropriate for experiences that may be less temporally defined, such as feeling sad.

Comparing Methods

Each of the methods described above has distinct advantages and disadvantages. Choosing a method requires the researcher to consider the nature of the stressor to be assessed and the research questions or hypotheses to be tested. There are also important logistical considerations in choosing a method. Sometimes, limitations of a given method can be overcome by combining different methods within the same study.

Interval-contingent methods are useful in documenting the amount of exposure an individual has to particular types of stressors or in assessing overall exposure levels in a given period of time. For example, a researcher may be interested in the number of arguments between spouses over a particular time period (e.g., a week, a month). As Stone et al. (1991) point out, the frequency with which records are kept (several times a day, once a day, once a week), as well as the total number of recording periods (e.g., 30 consecutive days, 6 consecutive months) should be based on an understanding of the stressor in question, the temporal relationship between the stressor and relevant outcomes, and a consideration of the statistical power needed for the analyses to be conducted. For example, as noted above, Bolger et al. (1989b) found that work conflicts were reported only about once every month. If a researcher was interested in the relationship between work conflict and mental health, it would clearly be inadequate to ask respondents to keep a daily diary for a total of one month, since this would yield an average of only one incident. However, a reliable estimate of instances of work overload, a much more common stressor for many people, may well be achieved with a daily diary kept for one month.

The nature of the outcome in question is also an important consideration in determining the length of time respondents should be asked to keep diaries. For example, if mood is the outcome of interest, daily recordings for 3 or 4 weeks will usually yield sufficient variability in mood scores to allow the investigator to explore the temporal relationship between a daily stressor such as work overload and changes in mood. However, rarer discrete outcomes such as absenteeism and utilization of health services would require that the study be lengthened to allow for enough of these events to occur.

There are two potential problems with interval-contingent methods. The most obvious is that respondents may be recording the data several hours or days after the event in question took place. This makes these methods susceptible to forgetting and biased recall (Suls & Martin, 1993; Wheeler & Reis, 1991). This problem is greater for continuous phenomena that vary considerably over relatively short time periods (e.g., mood, amount of perceived work stress, levels of concentration) than discrete events with a well-defined time of onset (e.g., argument, exercise, headache). Having respondents record information at more frequent intervals may help solve this problem, but may have the undesirable effect of overburdening the respondent.

Diary studies often instruct respondents to complete the diaries at the end of each day. An advantage to this practice is that respondents get accustomed to a regular routine of completing the diaries, making it less likely for them to forget to complete the record (although telephone reminders and daily collection of diaries help reduce this problem). It is also logical in terms of having respondents summarize their day. However, some research questions might dictate completion of the diaries at times other than the end of the day. For example, a work stress study may want respondents to complete the diary immediately after work; or a study investigating physiological changes related to myocardial ischemia may want recording to occur toward the end of the morning since there is evidence that there is a higher incidence of ischemia in the morning hours (Krantz et al., 1993). However, a potential problem with using any fixed time interval for recording daily, weekly, or, monthly data may occur if the phenomenon being measured tends to have a regular cyclical pattern. For example, more undesirable life events tend to be reported on weekdays, compared to weekends (Stone et al., 1985). As such, estimating the prevalence of such events based on a daily diary completed once a week for several weeks may be biased if always completed on either a weekday or a weekend.

Signal-contingent methods effectively eliminate the recall problem by having respondents immediately record their thoughts, feelings, or behaviors when they are signaled by the researcher. Since the respondents can be signaled at any time, this method may also eliminate the potential bias introduced by having the respondents complete the diaries at the same time of day or day of week. For assessing frequently occurring activities, as well as inner psychological and physiological states, signal-contingent methods would appear very useful. Studies using these methods have been particularly useful in documenting how various groups of people spend their daily lives, and as well as how inner psychological states co-vary with certain activities. For example, Csikszentmihalyi and Larson (1984) report that, among students, studying tends to be associated with negative mood, especially among low achievers.

There are at least two disadvantages to signal-contingent methods (Suls & Martin, 1993; Wheeler & Reis, 1991). First, because they are designed to assess immediate experience versus recall over a longer time period, they are impractical as a way to study infrequently occurring events or stressors. For example, if the research question concerns the relationship between arguments and physical symptoms, ESM-type methods would be unlikely to sample enough episodes of arguments to allow for a meaningful analysis. However, if the researcher is interested not in investigating a particular class of stressor, but, rather in assessing whether *any* stressor was occurring at that time, then ESM methods would be more feasible. Chronic stressors rather than acute stressors are more likely to be assessed with this method. This is especially true if samples of individuals susceptible to the chronic stressor in question are studied. For example, Williams et al. (1991) studied the relationship between juggling multiple roles and mood in a sample of working mothers using ESM methods.

The second major disadvantage of signal-contingent methods is that they are more intrusive than other methods since respondents have no control over when they will complete the records. As such, they may be signaled at times when it is inconvenient to fill out the data form (e.g., in class, at dinner, feeding the baby, in the shower, jogging). Some ESM studies allow respondents to turn off the signal devices when they do not want to be disturbed. For both reasons, respondents will not respond to a certain percentage of signals; thus some data will be missing. For example, Larson (1979), using an adolescent sample, reported that one-third of the signals were missed. Other studies with adults and children report a rate of missed signals of between 15 and 20 percent (e.g., Csikszentmihalyi & Graef, 1980; Larson & Lampman-Petraitis, 1989). Although we found no data on this issue, it seems logical to assume that the ability of respondents effectively to tolerate the intrusiveness of these methods would decrease as the level of disorganization and stress in their lives increased. If so, missing data may not be random in the sample studied. Pilot testing with high-risk samples is necessary to ensure that respondents can use these methods effectively.

Event-contingent methods, like signal-contingent methods, overcome problems with forgetting and selective recall by having respondents complete a record every time a particular type of event occurs. Unlike signal-contingent methods, these methods also can be used to study events with varying frequencies. The event that triggers the record could be either a predictor variable (e.g., stressful interpersonal interactions) or an outcome variable (e.g., asthma attacks, drug use). These methods are particularly useful for studying the immediate circumstances surrounding specific acute stressors or discrete outcomes. They are less feasible for studying chronic stressors since time of onset is not likely to be well defined (although some chronic stressors have an acute onset). Advances in the technology of physiological measurement will also expand the range of bodily conditions that could trigger a record (e.g., increased blood pressure or heart rate).

Event-contingent approaches require that respondents be well trained in the definition of the event to be studied. Most events involving social interaction, for example, will vary considerably in intensity and duration. For instance, a study of work stress may require respondents to complete a record every time they experience conflict with a supervisor. In such a study, clear instructions regarding the researcher's definition of "conflict" would need to be given to respondents to prevent them from being overly exclusive or inclusive. Even with extensive training, it is helpful to have respondents provide some open-ended descriptive information about the event so that the researcher can confirm that the incident being recorded meets the study definitions. An alternative approach, advocated by Stone, Kessler, & Haythornthwaite (1991), is to have interviewers who are trained to probe for relevant events make daily phone calls to respondents.

An interesting variant of the event-contingent approach was a study conducted by Bolger (Bolger, 1990; Bolger & Eckenrode, 1991) in which college students preparing for a medical school admission test kept daily diaries for $2^1/_2$ weeks before and $2^1/_2$ weeks after the exam. The stressful event, the exam, was therefore preselected by the researcher. The diary data could then be used to assess the students' psychological states and coping responses leading up to the exam. These data were particularly useful in documenting differences in the adjustment of the students to this stressor as a function of social integration and personality characteristics, measured prior to the diary period. This is an example of how diary methods can be used to study the impact of major as opposed to minor stressors. This approach is, of course, feasible only with events that can be anticipated.

Since they involve intensive self-monitoring of behavior over several occasions, all the methods described here are potentially reactive in nature. But this may be particularly true of event-contingent methods, depending on the amount of control the respondent has over the event in question. For example, if the event involves a behavior such as smoking or substance use, or a type of social interaction over which the respondent has some control, such as an argument with a spouse or child, completing a record every time such an event occurs may cause the behavior to change over time. This may contribute to the commonly observed pattern of response decay over time in diary studies (Stone et al., 1991; Verbrugge, 1980), although it is often difficult to determine whether a decrease in the rate of reporting events over time is due to fatigue, or to changes of feelings or behaviors resulting from the task. This uncertainty may be partly resolved by debriefing respondents after the diary-reporting period of the study is completed, asking them directly about the diary task and changes in their behavior over time. The diary period may also be shortened to prevent fatigue or sensitization. However, this may require a larger number of subjects to compensate for the loss of statistical power due to reducing the number of days.

Specific Daily Stress Measures

For each of the methods reviewed above, there are a variety of approaches to eliciting information about daily stressors, ranging from single open-ended questions (e.g., Eckenrode, 1984) to structured checklists (e.g., Stone & Neale, 1982).

The choice of approach depends on the focus of the study, the mode of data collection (e.g., paper-and-pencil diaries versus daily phone calls), and the circumstances and tolerance of the population being studied. We concentrate on checklist approaches here, although we first briefly discuss open-ended methods since they may be more suitable for some researchers.

An open-ended method was used by Eckenrode (1984) in a study of low-income women and children using a neighborhood health center in Boston (see also Caspi, Bolger, & Eckenrode, 1987; Gortmaker, Eckenrode, & Gore, 1982). In this study, these women were asked to complete a one-page diary at the end of each day for 28 days. The single question regarding stress was: "Did anything go wrong today in the house, with the children or others in the household, at work, or elsewhere?" If a respondent answered "yes," she were then asked to describe what happened. Such responses could then be classified into discrete categories. This diary was a modification of a diary used by Roghmann and Haggerty (1972), who asked three separate open-ended questions, beginning with the stem question "Did anything go wrong today?" and then varying the domain ("at work or in the house"; "with the school or the children"; "getting along with friends, relatives, husband"). More recently, Emmons (1991) had undergraduate subjects keep a daily diary (completed twice a day) in which they were asked to list four events or thoughts, two that contributed to negative mood and two to positive mood. Coders then categorized events into categories (e.g., achievement, interpersonal).

As Stone et al. (1991) point out, there are both advantages and disadvantages to such an open-ended approach. Simple, open-ended questions are easy for respondents to understand and to complete in little time. This may be attractive in studies where stress is not the major focus of the study and where time is limited. If phone interviews are used rather than paper-and-pencil diaries, open-ended questions can also establish a more conversational style and greater rapport with respondents. If combined with probes for various domains of stress, the resulting data can then be coded by the researcher into discrete categories of stressful events.

There are also several disadvantages to open-ended approaches. In paper-and pencil versions, they require respondents to write out a description of the events. The amount of detail provided by respondents will vary greatly, leading to potential problems in classifying responses. Recall is also more likely to be biased by personality traits or transient mood states when an open-ended versus a structured format is used, although for minor stressors, the amount of recall error with diaries will generally be much lower than with retrospective interviews. Finally, open-ended questions will tend to underestimate the total exposure of respondents to stressors on a given day. If the researcher is interested in knowing only whether *any* stressor occurred that day, these methods may be adequate. However, if the research question requires a more complete assessment of all stressors occurring that day, other approaches, such as the checklist methods described below, are more appropriate.

As an alternative to open-ended questions about stressful events, some studies have used brief global ratings of perceived stress levels. For example, in a study of the relationship between daily stress and recurrence of genital herpes simplex, Rand and colleagues (1990) had respondents complete a Daily Stress Questionnaire which asked them to indicate on a 4-point Likert scale how much stress they felt that day in six areas (e.g., physical health, financial, relations with family). Watson (1988), in a daily study of the correlates of negative and positive affect, asked college students, "How much stress (e.g., because of hassles, demands) were you under today?" Responses were recorded on a 5-point Likert scale (1 = felt very slightly or not at all; 5 = felt very much). Such approaches may provide convenient summary measures of perceived stress at the daily level, and may be a practical alternative to more comprehensive assessments of daily events when the amount of space devoted to stress assessment in the diary is very limited. However, these approaches may sample only a small portion of potentially stressful experiences, or measure stress at such a global level so as to preclude analyses that differentiate between types of stressors. Global assessments of perceived stress also tend to confound the objective exposure to stressors with the respondent's reaction to them (e.g., distress), a problem similar to that faced by the assessment of major life events (see chapters 2 and 3).

Most studies where stress is a major focus of interest now use some form of checklist. Researchers sometimes supplement the checklist with open-ended questions to assess events that may have been missed with the checklist, and to obtain more detailed contextual information about checked events. In the next section we review a sample of checklist measures. Although our survey of checklist measures is not a comprehensive list, it is designed to represent the diversity of approaches in current use.

Daily Life Experience (DLE) checklist

This checklist was originally named the Assessment of Daily Experience (ADE), with its development described by Stone and Neale (1982). Its current version is called the Daily Life Experience checklist (DLE).[1] It contains 78 events organized into five major domains (i.e., Work; Leisure; Family and friends; Financial; Other). Respondents are asked to rate those events that "happened since you first awoke this morning." For events that are checked, respondents also rate the desirability of the event on a 6-point scale, from "extremely undesirable" to "extremely desirable"; as well as the meaningfulness of the event on a 3-point scale from slightly to extremely meaningful. Events generally rated as positive (e.g., promotion or raise) as well as negative (e.g., argument with spouse) are included. Respondents can also write in any events not on the list, and are asked to write in the "worst or most bothersome problem of the day." Typically the diary also contains mood and symptom items. The diary takes approximately 10 minutes to complete.

A desirable feature of this instrument is that events were sampled by having persons in the population to be studied nominate, in an open-ended format, events that occurred to them over 14 days. The large number of generated events were then summarized into major and minor categories, with the resulting structured checklist used in subsequent diary studies. This procedure ensured that events were representative of the population from which research samples would be drawn.

Another distinctive feature of this research program is the researchers' use of couples as part of a "target–observer" procedure. Married couples were recruited from the community, and each was asked to complete diaries for the same days. Husbands were the target respondents; the wives were the observers, filling out the forms in terms of events that had occurred to the husbands. Each would fill out the forms independently, then meet to discuss and revise their lists. In this way, many events initially overlooked by the respondents (husbands) could be added to the list, yielding a more reliable daily score. Degree of concordance between partners was also assessed and used as a method to assess the validity of the data collected with the instrument (see Stone & Neale, 1982).

The Hassles Scale

The original Hassles Scale (DeLongis et al., 1982; Kanner et al., 1981) consists of a list of 117 items generated by the research staff covering the areas of work, health, family, friends, the environment, practical considerations, and chance occurrences. Respondents also rated the severity of items on a 3-point scale (somewhat, moderately, extremely), yielding a frequency score (count of number of hassles checked), and an intensity score (mean severity rating). Although developed as a measure of everyday stressors, the scale was originally intended not for daily use, but rather, for once-a-month administration for several consecutive months. A companion Uplifts Scale was also developed by these researchers, consisting of 135 items reflecting positive experiences in areas similar to those in the Hassles Scale.

The original research with 100 community residents of Alameda County, California clearly points out that such checklists frequently tap into the everyday manifestations of chronic stressors as well as discrete microstressors not necessarily tied to enduring difficulties. The five most frequently cited hassles in this sample were "concerns about weight," "health of a family member," "rising prices of common goods," "home maintenance," and "too many things to do." Although the distinctiveness of daily stress measures from more aggregate measure of chronic stress is an important theoretical and methodological issue (see Chapter 5; Wheaton, in press), the issue that has resulted in more controversy regarding the Hassles Scale concerns possible confounding of items with measures of psychological distress or psychopathology (Dohrenwend, Dohrenwend, Dodson, & Shrout, 1985; Dohrenwend & Shrout, 1985; Lazarus, DeLongis, Folkman, & Gruen, 1985; Monroe, 1983). Some items, such as "you have been lonely," "you have troubling thoughts about your future," seem less focused on environmental stressors than on internal states.

In a study of 75 married couples, DeLongis, Folkman, and Lazarus (1988) report on a revised 53-item Hassles and Uplifts Scale that attempts to remove items that are redundant or that suggest psychological or somatic symptoms. For each item respondents rate on 4-point scales the extent to which it was a hassle and an uplift, allowing items to contain both qualities. In this study the scale was completed for 4-day periods between each of six monthly interviews, resulting in 20 daily assessments. This scale appears to overcome some of the conceptual and methodological limitations of the original scale and is short enough to be practical for daily use.

Other investigators have modified the Hassles Scale for specific research questions or populations. For example, Lepore, Evans, and Palsane (1991), in a study of males in urban India, constructed a 5-item "social hassles" scale from items in the Hassles Scale, asking respondents to report how often they experienced these within the last 2 months. Wheaton (in press) recently used an abbreviated daily hassles measure composed of 25 items from the original Hassles Scale within an interview format as part of a community survey. As with measures of major life events, such as the Schedule of Recent Experiences (SRE: Holmes & Rahe, 1967), researchers will continue to adapt such measures to address specific research questions or meet the unique needs of their research designs.

Additional measures modeled after the original Hassles Scale have begun to appear in the literature even if their component items are not drawn directly from it. For example, the Parenting Daily Hassles Scale (Crnic & Booth, 1991; Crnic & Greenberg, 1990) is a 20-item scale used in a paper-and-pencil questionnaire format in which parents report on the frequency and intensity of everyday events in parenting and parent–child relationships (e.g., continually cleaning up kids' messes; being nagged, whined at, or complained to). These studies have used a 6-month recall period, although the investigators also claim having used the scale covering "the past several days." To date, this scale has not been used as part of a daily diary methodology, but with minor modifications seems appropriate in such studies.

Bolger et al. Daily Stress Scale

A 22-item daily event scale has been reported as part of a daily diary study involving 166 married couples drawn from a larger study of 778 couples in the Detroit metropolitan area (Bolger & Schilling, 1991; Bolger et al., 1989a, 1989b). The 22 events were selected on the basis of earlier pilot testing with a sample of 64 married couples where an open-ended format was used to identify common daily stressors in this population. Events selected for the final scale occurred on at least 5 percent of the days and were shown to be associated with distressed daily mood in the pilot sample. Respondents are asked to check whether the event happened to them over the preceding 24 hours.

In one study (Bolger et al., 1989a), the 22 events were aggregated into 10 summary event categories (e.g., overload at home, family demands, interpersonal conflicts or tensions with one's spouse), with each category being represented as a dummy variable in a regression analysis. In another study (Bolger et al., 1989b), 7 of the items were examined, involving overloads or interpersonal tensions or arguments since the research questions focused on the contagion of stress across work and family roles.

Because this study had a particular interest in interpersonal tensions or arguments, respondents were also asked to identify the most serious event of this kind on each day. A series of follow-up questions were then asked to provide more detailed information about that event (e.g., felt responsibility; amount of perceived control; who was helpful or made the situation worse; coping responses). This is a good example of how a diary method can be used to assess generalized exposure to minor stressors while also gathering detailed information on a specific class of stressor. Of course, individual researchers could modify such a diary instrument to focus attention on any of several categories of events.

Inventory of Small Life Events (ISLE)

This inventory contains 178 items, 98 undesirable and 80 desirable events (see Zautra, Guarnaccia, & Dohrenwend, 1986, for a complete list). Similar to the Hassles Scale, this inventory is designed for use within an interview format with a 1-month recall period. Its length makes it impractical for use in most daily diary studies.

The authors claim that this scale addresses some of the problems of other checklists by (1) excluding items that could be confounded with psychological states; (2) distinguishing events from ongoing or habitual activities by defining them as changes from usual day-to-day occurrences; (3) distinguishing on an a priori basis desirable from undesirable events; (4) including only observable events; and (5) distinguishing small events from major life events. In the latter respect, the inventory is designed to complement the PERI life event scale (B. S. Dohrenwend, Krasnoff, Askenasy, & Dohrenwend, 1978).

Events were selected from previous scales, with items modified and new items added based on input from other researchers and from pilot studies. As part of the scale's development, expert judges and students provided readjustment ratings, using procedures similar to those employed with the PERI. A cutoff score of 250 was used to distinguish small from large events (for each category of events, subjects were provided an anchor event from the PERI). Desirability ratings also were collected with these samples to verify the initial classifications of events. Scores are based on counts across the entire inventory or within categories of events.

A modified version of the ISLE, with items rewritten to reflect events appropriate for older adults is reported in a study by Zautra and associates (Zautra, Finch, Reich, & Guarnaccia, 1991; Zautra, Reich, & Guarnaccia, 1990). As with the original inventory, experts were used to rate the degree of readjustment required by the events as well as event desirability. Respondents completed the inventory once a month for 10 months, a procedure that allows for the identification of recurring events. For example, whereas the event "criticized by spouse/mate" was reported by only 8.8 percent of the sample in the first interview, for those who reported the event there was a 26 percent chance of recurrence in the subsequent month.

Daily Stress Inventory

The Daily Stress Inventory (DSI) is a 58-item questionnaire designed to assess the sources and magnitude of minor stressful events (Brantley & Jones, 1989, 1993). A manual is available describing the development of the inventory, its administration, and scoring information, as well as normative data from adult, student, and medical patient samples.

Items were selected for the scale based on responses to open-ended questions regarding stressful events in diaries completed over a 2-week period by samples of psychology graduate students, community volunteers, and persons seeking counseling services. The item pool was further refined in a community study of 433 adults. Items are grouped into six categories: interpersonal problems (e.g., "argued with another person"), personal competency (e.g., "was late for work/appointment"); cognitive stressors (e.g., "heard some bad news"); environmental hassles (e.g., "had minor accident"); and varied stressors (e.g., "misplaced something"). Respondents read through the list and rated those events that had occurred on a 7-point scale ranging from 1 = "occurred but was not stressful" to 7 = "caused me to panic." No explanation is given for why extreme perceived stress is characterized by "panic," rather than other psychological states such as depression. This appears to be a limiting feature of this instrument.

An attempt was made to eliminate items overtly referring to psychiatric or physical illness. Scores representing the total number of events checked and their impact (based on summated rating scores) have been used in studies by these researchers. The manual presents reliability and validity data (see also Brantley, Cocke, Jones, & Goreczny, 1988). To date, the measure has been used mainly in studies examining physiological responses to stress (e.g., Brantley, Dietz, McKnight, Jones, & Tully, 1988) and daily fluctuations in symptoms associated with psychosomatic disorders such as asthma (e.g., Goreczny, Brantley, Buss, & Waters, 1988).

Future Directions

Our review has provided an introduction to existing measures of daily stressors and some guidelines for choosing among these measures. In this final section we briefly discuss some promising future directions in the measurement of daily stress.

Checklist measures of major life events have a number of significant limitations relating to the reliability and validity of the event categories. The work of Brown and others has demonstrated that the measurement of life events can be significantly improved by use of in-depth interview measures, such as the Life Events and Difficulties Schedule (LEDS), that involve intensive probes for the degree of threat posed by each event (see Chapter 3). Checklist measures of daily events are prone to many of the same limitations as are checklist measures of major stressors. In principle, then, the measurement of daily stress could be improved upon by use of an interview approach. Obviously, a face-to-face interview would be impractical for daily stress assessment, but a nightly telephone interview might be feasible. To our knowledge, though, no work is currently underway to adapt interview-based methods to the study of daily stressors. (See Stone et al., 1991, for a discussion of the potential advantages of phone interviews in diary studies.)

In studies of daily stress, it is customary to have subjects complete measures at least at daily intervals, although sometimes the unit of analysis is less frequent than that. Yet, arguably, the unit of analysis in daily event studies should be smaller than a single day. For example, a daily event that occurs in the morning is likely have its greatest effect soon after it occurs and may have no detectable (residual) effect by bedtime. Yet many daily event measures are not suitable for frequent use during a single day, and none of the daily event measures reviewed above has been specifically tailored for such use. Stone, Neale, and Shiffman (1993) have recently pointed out that lagged or prospective effects of daily stressors have rarely been found in the literature. It is conceivable that measures of daily stress and mood taken at sufficiently frequent intervals would show prospective effects. Such measures, of course, would need to be considerably shorter than most existing daily stress measures.

One potentially promising innovation for daily stress measurement has been the development of ambulatory mood-measurement instruments. Shiffman and his colleagues have had subjects use small hand-held recording devices to make simple ratings of moods at many points during a day (Shiffman, Paty, Gnys, & Kassel, 1992). As yet, these devices have not been used to measure daily events—and of necessity the number of event categories used would need to be quite small—but this seems to be a very fruitful area for further research.

To date, most daily stress research has either ignored event content or attempted to cover a wide range of event content. What is emerging, however, is that not all types of daily events are equally stressful. For example, Bolger and colleagues (1989a) found that adult interpersonal conflicts were by far the most stressful types of daily events in their sample. This observation suggests the usefulness of developing daily stress measures that focus on specific types of events and that probe into the characteristics of these events in depth. Few such measures currently exist. Furthermore, it may be valuable to use an event-contingent measurement strategy to record exposure to particular events such as daily conflicts, a strategy where subjects complete a very short diary each time an event of a certain type occurs. In this way, researchers could obtain more fine-grained information about the characteristics and contexts of specific classes of stressors.

One neglected issue in the study of daily events is the likely reactivity of daily event measures when they are used in a daily diary study. For example, event frequency declines as the number of days of diary recording increases (Bolger et al., 1989a). Also, distress is higher during the first few days of a diary study than on later days (Bolger, 1990). There are two possible solutions to these problems. First, where the research question of interest is the association between events and outcomes, day can be used as a control variable in the statistical analyses. Second, it may be feasible to discard the first few days' data if these seem to be strongly affected by subject reactivity (Tennen, Suls, & Affleck, 1991).

Finally, the issue of cultural, racial, ethnic, or social class variability in the experience of daily stress has not been adequately addressed by research to date. Although several daily stress measures described above were developed with community samples, the reported analyses generally do not focus on social group differences in daily stress processes. There is also little information in the literature regarding the feasibility of conducting daily stress studies with children.

It is clear that daily diary studies provide a window into the stress process that more traditional interview and questionnaire methods cannot hope to achieve. On the other hand, these methods are labor- and cost-intensive, and as such, the decision to use these measures should be taken with care. The researcher using daily assessment methods is faced with a number of difficult decisions (see Stone et al., 1991), the most important being whether measuring stress at a daily level can yield data that can address important theoretical issues. As with other areas of measurement reviewed in this volume, there is no one best measure to fit all research needs. Rather than recommend specific measures, we have attempted to provide the researcher with an overview of this emerging area of research and the measurement approaches currently found in the literature.

Note

1. Information regarding this scale can be obtained from Dr. Arthur Stone, Department of Psychiatry and Behavioral Science, School of Medicine, SUNY at Stonybrook, Putnam Hall, South Campus, Stonybrook, NY 11794.

References

Bolger, N. (1990). Coping as a personality process: A prospective study. *Journal of Personality and Social Psychology, 59,* 525–537.

Bolger, N., DeLongis, A., Kessler, R. C., & Schilling, E. A. (1989a). Effects of daily stress on negative mood. *Journal of Personality and Social Psychology, 57,* 808–818.

Bolger, N., DeLongis, A., Kessler, R. C., & Wethington, E. (1989b). The contagion of stress across multiple roles. *Journal of Marriage and the Family, 51,* 175–183.

Bolger, N., & Eckenrode, J. (1991). Social relationships, personality, and anxiety during a major stressful event. *Journal of Personality and Social Psychology, 61,* 440–449.

Bolger, N., & Schilling, E. A. (1991). Personality and the problems of everyday life: The role of neuroticism in exposure and reactivity to daily stressors. *Journal of Personality, 59,* 355–386.

Brantley, P. J., Cocke, T. B., Jones, G. N., & Goreczny, A. J. (1988). The Daily Stress Inventory: Validity and effect of repeated administration. *Journal of Psychopathology and Behavioral Assessment, 10,* 75–81.

Brantley, P. J., Dietz, L. S., McKnight, G. T., Jones, G. N., & Tulley, R. (1988). Convergence between the Daily Stress Inventory and endocrine measures. *Journal of Consulting and Clinical Psychology, 56,* 549–551.

Brantley, P. J., & Jones, G. N. (1989). *The Daily Stress Inventory: Professional manual.* Odessa, FL: Psychological Assessment Resources.

Brantley, P. J., & Jones, G. N. (1993). Daily stress and stress-related disorders. *Annals of Behavioral Medicine, 15,* 17–25.

Bryk, A. S., & Raudenbush, S. W. (1992). *Hierarchical linear models: Applications and data analysis methods.* Newbury Park, CA: Sage.

Campbell, J. D., Chew, B., & Scratchley, L. S. (1991). Cognitive and emotional reactions to daily events: The effects of self-esteem and self-complexity. *Journal of Personality, 59,* 473–505.

Caspi, A., Bolger, N., & Eckenrode, J. (1987). Linking person and context in the daily stress process. *Journal of Personality and Social Psychology, 52,* 184–195.

Crnic, K. A., & Booth, C. L. (1991). Mother's and father's perceptions of daily hassles of parenting across early childhood. *Journal of Marriage and the Family, 53,* 1042–1050.

Crnic, K. A., & Greenberg, A. S. (1990). Minor parenting stress with young children. *Child Development, 61,* 209–217.

Csikszentmihalyi, M., & Graef, R. (1980). The experience of freedom in daily life. *American Journal of Community Psychology, 8,* 401–414.

Csikszentmihalyi, M., & Larson, R. (1984). *Being adolescent: Conflict and growth in the teenage years.* New York: Cambridge University Press.

Csikszentmihalyi, M., Larson, R. (1987). Validity and reliability of experience sampling method. *Journal of Nervous and Mental Diseases, 175,* 526–536.

Csikszentmihalyi, M., Larson, R., & Prescott, S. (1977). The ecology of adolescent experience. *Journal of Youth and Adolescence, 6,* 281–294.

DeLongis, A., Coyne, J., Dakof, G., Folkman, S., & Lazarus, R. S. (1982). The relationship of hassles, uplifts, and major life events to health status. *Health Psychology, 1,* 119–136.

DeLongis, A., Folkman, S., & Lazarus, R. S. (1988). The impact of daily stress on health and mood: Psychological and social resources as mediators. *Journal of Personality and Social Psychology, 54,* 486–495.

Dohrenwend, B. S., Dohrenwend, B. P., Dodson, M., & Shrout, P. E. (1984). Symptoms, hassles, social supports, and life events: Problem of confounded measures. *Journal of Abnormal Psychology, 93,* 222–230.

Dohrenwend, B. S., Krasnoff, L., Askenasy, A. R., & Dohrenwend, B. P. (1978). Exemplification of a method for scaling life events: The PERI Life Events Scale. *Journal of Health and Social behavior, 19,* 205–229.

Dohrenwend, B. P., & Shrout, P. E. (1985). "Hassles" in the conceptualization and measurement of life stress variables. *American Psychologist, 40,* 780–785.

Downes, J., & Collins, S. D. (1940). A study of illness among families in the Eastern Health District of Baltimore. *Milbank Memorial Fund Quarterly, 18,* 5–26.

Duck, S. W. (1991). Diaries and logs. In B. M. Montgomery & S. W. Duck (Eds.), *Studying social interaction* (pp. 141–161). New York: Guilford.

Duncan, A. D. (1969). Self-application of behavior modification techniques by teen-agers. *Adolescence, 16,* 541–556.

Eckenrode, J. (1984). The impact of chronic and acute stressors on daily reports of mood. *Journal of Personality and Social Psychology, 46,* 907–918.

Emmons, R. A. (1991). Personal strivings, daily life events, and psychological and physical well-being. *Journal of Personality, 59,* 453–472.

Epstein, S. (1979). The stability of behavior: I. On predicting most of the people much of the time. *Journal of Personality and Social Psychology, 37,* 1097–1126.

Epstein, S. (1980). The stability of behavior: II. Implications for psychological research. *American Psychologist, 35,* 790–806.

Goreczny, A. J., Brantley, P. J., Buss, R. R., & Waters, W. F. (1988). Daily stress and anxiety and their relation to daily fluctuations of symptoms in asthma and chronic obstructive pulmonary disease (COPD) patients. *Journal of Psychopathology and Behavioral Assessment, 10,* 259–267.

Gortmaker, S., Eckenrode, J., & Gore, S. (1982). Stress and the utilization of health services: A time-series and cross-sectional analysis. *Journal of Health and Social Behavior, 23* (I), 25–38.

Hedeker, D. (1993). MIXREG. A fortran program for mixed-effects linear regression models. Chicago: University of Illinois, Prevention Research Center.

Holmes, T. H., & Rahe, R. H. (1967). The Social Readjustment Rating Scale. *Journal of Psychosomatic Research, 11,* 213–218.

Hormuth, S. E. (1986). The sampling of experiences in situ. *Journal of Personality, 54,* 262–293.

Jaccard, J., & Wan, C. K. (1993). Statistical analysis of temporal data with many observations: Issues for behavioral science data. *Annals of Behavioral Medicine, 15,* 41–50.

Kanner, A. D., Coyne, J. C., Schaefer, C., & Lazarus, R. S. (1981). Comparison of two modes of stress measurement: Daily hassles and uplifts versus major life events. *Journal of Behavioral Medicine, 4,* 1–39.

Kenny, D. A., Kashy, D. A., & Bolger, N. (1993). The analysis of multi-level data. Unpublished manuscript, Department of Psychology, University of Connecticut.

Kessler, R. C., Price, R. H., & Wortman, C. B. (1985). Social factors in psychopathology: Stress, social support, and coping processes. *Annual Review of Psychology, 36,* 531–572.

Krantz, D. S., Gabbay, F. H., Hedges, S. M., Leach, S. G., Gottdiener, J. S., & Rozanski, A. (1993). Mental and physical triggers of silent myocardial ischemia: Ambulatory studies using self-monitoring diary methodology. *Annals of Behavioral Medicine, 15,* 33–40.

Larsen, R. J., & Kasimatis, M. (1991). Day-to-day physical symptoms: Individual differences in the occurrence, duration, and emotional concomitants of minor daily illnesses. *Journal of Personality, 59,* 387–423.

Larson, R. (1979). The significance of solitude in adolescent lives. Unpublished doctoral dissertation, University of Chicago, 1979.

Larson, R., & Csikszentmihalyi, M. (1983). The experience sampling method. In H. T. Reis (Ed.), *Naturalistic approaches to studying social interaction.* San Francisco: Jossey-Bass.

Larson, R., & Ham, M. (1993). Stress and "stress and storm" in early adolescence: The relationship of negative events with dysphoric affect. *Developmental Psychology, 29,* 130–140.

Larson, R., & Lampman-Petraitis, C. (1989). Daily emotional stress as reported by children and adolescents. *Child Development, 60,* 1250–1260.

Lazarus, R. S., DeLongis, A., Folkman, S., & Gruen, R. (1985). Stress and adaptational outcomes: The problem of confounded measures. *American Psychologist, 40,* 770–779.

Lepore, S. J., Evans, G. W., & Palsane, M. N. (1991). Social hassles and psychological health in the context of chronic crowding. *Journal of Health and Social Behavior, 32,* 357–367.

Lewinsohn, P. M., & Libet, J. (1972). Pleasant events, activity schedules, and depression. *Journal of Abnormal Psychology, 79,* 291–295.

Lewinsohn, P. M., & Talkington, J. (1979). Studies on the measurement of unpleasant events and relations with depression. *Applied Psychological Measurement, 3,* 83–101.

Longford, N. T. (1986). VARCL—Interactive software for variance component analysis. *Professional Statistician, 74,* 817–827.

Mason, W. M., Anderson, A. F., & Hayat, N. (1988). Manual for GENMOD. Population Studies Center, University of Michigan.

Meyer, R. J., & Haggerty, R. J., (1962). Streptococcal infections in families: Factors altering individual susceptibility. *Pediatrics, 29,* 539–549.

Monroe, S. M. (1983). Major and minor life events as predictors of psychological distress: Further issues and findings. *Journal of Behavioral Medicine, 6,* 189–205.

Nowlis, V. (1965). Research with the mood adjective check list. In S. Tomkins & C. Izard (Eds.), *Affect: Measurement of awareness and performance* (pp. 352–389). New York: Springer-Verlag.

Pearlin, L. I., Lieberman, M. A., Menaghan, E., & Mullan, J. T. (1981). The stress process. *Journal of Health and Social Behavior, 22,* 337–356.

Prosser, R., Rasbash, J., & Goldstein, H. (1991). *ML3 software for three-level analysis: User's guide for V. 2.* London, UK: Institute of Education, University of London.

Rand, K., Hoon, E., Massey, J., & Johnson, J. (1990). Daily stress and the recurrence of genital herpes simplex. *Archives of Internal Medicine, 150,* 1889–1893.

Repetti, R. L. (1989). Effects of daily workload on subsequent behavior during marital interaction: The roles of social withdrawal and spouse support. *Journal of Personality and Social Psychology, 57,* 651–659.

Roghmann, K., & Haggerty, R. J. (1972). The diary as a research instrument in the study of health and health behavior, *Medical Care, 10,* 143–163.

Roghmann, K., & Haggerty, R. J. (1973). Daily stress and the use of health services in young families. *Pediatrics Research, 7,* 520–526.

Shiffman, S., Paty, J., Gnys, M., & Kassel, J. (1992). Computerized self-monitoring with base-rate control: Methodological analyses. In S. Shiffman (chair), *Field assessment methods: Studying behavior in its natural environment.* Thirteenth Annual Meeting of the Society of Behavioral Medicine. New York: March 25–28, 1992.

Stone, A. A. (1987). Event content in a daily survey differentially predicts mood. *Journal of Personality and Social Psychology, 52,* 56–58.

Stone, A. A., Hedges, S. M., Neale, J. M., & Satin, M. S. (1985). Prospective and cross-sectional mood reports after no evidence of a "Blue Monday" phenomenon. *Journal of Personality and Social Psychology, 49,* 129–134.

Stone, A. A., Kessler, R. C., & Haythornthwaite, J. A. (1991). Measuring daily events and experiences: Decisions for the researcher. *Journal of Personality, 59,* 575–605.

Stone, A. A., & Neale, J. M. (1982). Development of a methodology for assessing daily experiences. In A. Baum & J. E. Singer (Eds.), *Advances in environmental psychology: Environment and health* (Vol. 4, pp. 49–83). Hillsdale, NJ: Lawrence Erlbaum.

Stone, A. A., Neale, J. M., & Shiffman, S. (1993). Daily assessments of stress and coping and their association with mood. *Annals of Behavioral Medicine, 15*(1), 8–16.

Stone, A. A., & Shiffman, S. (1992). Reflections on the intensive measurement of stress, coping, and mood, with an emphasis on daily measures. *Psychology and Health, 7,* 115–129.

Sudman, S., & Ferber, R. (1971). Experiments in obtaining consumer expenditures by diary methods. *Journal of the American Statistical Association, 66,* 725–735.

Suls, J. S., & Martin, R. E. (1993). Daily recording and ambulatory monitoring methodologies in behavioral medicine. *Annals of Behavioral Medicine, 15,* 3–7.

Szalai, A. (Ed.). (1972). *The use of time: Daily activities of urban and suburban populations in twelve countries.* The Hague: Mouton.

Tennen, H., Suls, J., & Affleck, G. (1991). Personality and daily experience: The promise and the challenge. *Journal of Personality, 59,* 313–337.

Verbrugge, L. (1980). Health diaries. *Medical Care, 18,* 73–95.

Verbrugge, L. (1985). Triggers of symptoms and health care. *Social Science and Medicine, 20,* 855–876.

Watson, D. (1988). Intraindividual and interindividual analyses of positive and negative affect: Their relation to health complaints, perceived stress, and daily activities. *Journal of Personality and Social Psychology, 54,* 1020–1030.

West, S. G., & Hepworth, J. T. (1991). Statistical issues in the study of temporal data: Daily experiences. *Journal of Personality, 59,* 609–662.

Wheaton, B. (in press). Sampling the stress universe. In W. R. Avison & I. H. Gotlib (Eds.), *Stress and mental health: Contemporary issues and prospects for the future.* New York: Plenum Publishing.

Wheeler, L., & Nezlek, J. (1977). Sex differences in social participation. *Journal of Personality and Social Psychology, 35,* 742–754.

Wheeler, L., & Reis, H. T. (1991). Self-recording of everyday life events: Origins, types, and uses. *Journal of Personality, 59,* 339–354.

Williams, K. J., Suls, J., Alliger, G. M., Learner, S. M., & Wan, C. K. (1991). Multiple role juggling and daily mood states in working mothers: An experience sampling method. *Journal of Applied Psychology, 76,* 664–674.

Wong, M., & Csikszentmihalyi, M. (1991). Motivation and academic achievement: The effects of personality traits and the quality of experience. *Journal of Personality, 59,* 539–573.

Zautra, A. J., Finch, J. F., Reich, J. W., & Guarnaccia, C. A. (1991) Predicting the everyday events of older adults. *Journal of Personality, 59,* 507–538.

Zautra, A. J., Guarnaccia, C. A., & Dohrenwend, B. P. (1986). Measuring small life events. *American Journal of Community Psychology, 14,* 629–655.

Zautra, A. J., Reich, J. W., & Guarnaccia, C. A. (1990). Some everyday life consequences of disability and bereavement for older adults. *Journal of Personality and Social Psychology, 57,* 550–561.

5

Measurement of Chronic Stressors

Stephen J. Lepore

There are many reasons why investigators may wish to identify and measure chronic stressors, or distinguish chronic stressors from more episodic ones. One reason is that theoretically it is more plausible to link diseases that have a long-term development or onset phase, such as cardiovascular diseases, to stressors that are persistent and repeated rather than time-limited and episodic (cf. House, 1987; Krantz, Contrada, Hill, & Friedler, 1988; Chapter 1, this volume). Another reason is that there is emerging evidence that the human body does not always habituate to chronic stressors. For instance, in a recent meta-analysis of the literature on stress and immunity, Herbert and Cohen (1993) found that stressors lasting more than a month (e.g., caregiving, unemployment, bereavement) exerted a negative effect on multiple immune-system parameters in humans. These findings are consistent with theoretical notions that there are biological costs associated with adaptation to chronic stressors. People who are chronically exposed to stressors may exhaust all of their biological resources for adapting to stressors (Selye, 1956), or they might experience a persistently heightened state of arousal and associated bodily derangements that are potentially unhealthy (Cannon, 1929).

In addition to the direct negative effects of chronic stressors on the body, there could be indirect, or secondary, effects that result from individuals' efforts to cope (Cohen, Evans, Stokols, & Krantz, 1986; Lepore & Evans, in press). According to the cost of coping model (Cohen et al., 1986), coping efforts that are designed to remove stressors, or to alleviate distress associated with stressors, can have unintended negative health effects. Prolonged and active coping with a stressor can be psychologically fatiguing, resulting in attention, thought, and performance deficits (Cohen, 1980) that could make people prone to accidents. Persistent efforts to control or avert a stressor could result in increased neuroendocrine production and hemodynamic activity that might influence the onset or maintenance of diseases, especially coronary and cardiovascular diseases (Jenkins, 1988). Moreover, repeated and failed attempts to cope with a chronic stressor could induce a sense of helplessness and concomitant psychological distress (Seligman, 1975). Self-destructive and escapist forms of coping (e.g., alcohol and drug use) could have harmful bodily effects if they become routine and excessive—as they might in the face of chronic stressors. Finally, some chronic stressors can cause people to become socially avoidant, thereby diminishing their social support resources, which are potentially beneficial to health (Lepore, Evans, & Schneider, 1991). All of these secondary effects of exposure to a chronic stressor may pose threats to health that are equal to or even greater than the direct effects of the chronic stressor (Lepore & Evans, in press).

Emerging evidence also suggests that people do not psychologically habituate to chronic stressors. For example, stressors lasting for a year or longer have been shown to predict higher levels of psychological distress symptoms in married couples (McGonagle & Kessler, 1990) and in disabled and nondisabled adults (Avison & Turner, 1988). In addition, chronic stressors can heighten the unfavorable implications of acute stressors, exacerbating their effects. For example, Brown and Harris (1978) observed that women who had a pregnancy in a year in which they were experiencing chronic stressors (e.g., grossly inadequate housing) were more likely to be clinically depressed than women who had a pregnancy without a concurrent chronic stressor. Similarly, Lepore, Evans, and Schneider (1992) found that acute social stressors (e.g., arguments with housemates) were associated with psychological distress symptoms in people who lived in crowded households but not in people living in relatively uncrowded households. These data suggest that chronic stressors may increase psychological vulnerability to acute stressors by stripping people of their biological, psychological, and social coping resources.

Despite the compelling preliminary evidence that chronic stressors have unique effects on physical and psychological health, and despite the pervasiveness of chronic stressors in everyday life (Veroff, Douvan, & Kukla, 1981), there have been relatively few studies that have carefully measured chronic stressors and examined their effects on health. Our ignorance about chronic stressors can possibly be attributed to stress researchers' great reliance on life-event checklist measures (see Turner & Wheaton, Chapter 2, this volume). Researchers using life-event checklists derive a cumulative stress score that is based either on the number of events experienced within a specified time period (e.g., 6–12 months) or on the sum of the events' weights, which are based on judges' ratings of the amount of readjustment required by each event. This approach gives no direct information on the duration of the cumulated stressors, even though there can be much intrastressor and interstressor variability in the duration of stressors listed on life-event checklists. Thus, two respondents who only tick "job loss" on a life-events checklist would receive the same stress score, even if one of the respondents had been unemployed for 6 days and the other for 6 months. The loss of interstressor and intrastressor variability across subjects that results from ignoring duration could contribute to the generally low predictive power of checklist measures (cf. Avison & Turner, 1988), as well as obscure the unique processes through which stressors of different durations might affect health-related outcomes.

Chronic stressors are usually conceived of as discrete events and conditions, or constellations of related events and conditions, that persist over time. The preferred method of measuring them is one that is free of subjective biases, or error, and produces consistent results when used by different investigators or in different populations. A number of discrete stressors can be measured according to these criteria; these include environmental stressors such as crowding, noise, and air pollution (Evans & Cohen, 1987), and economic stressors such as unemployment or inability to pay bills (Dooley & Catalano, 1988). Investigators who focus on a discrete environmental or economic stressor often can obtain detailed objective information about the stressor's duration and magnitude by using archival records and electronic recording devices. For example, noise pollution can be precisely measured using a sound-level meter. Multiple sound readings recorded and averaged over time can provide a reliable index of chronic noise exposure. Chronic urban stressors, such as population density and crime rates, can be estimated using official archives that are generated annually by state agencies. Financial difficulties can be estimated by examining subjects' financial records, particularly tax documents and expense records.

Unfortunately, not all stressors are readily measured using objective archival materials or electronic recording devices. Therefore, researchers often must use somewhat more traditional and subjective social-science techniques, which involve the self-report observations of the subject (e.g., questionnaires, interviews) or of another human observer (e.g., naturalistic observation, laboratory observation, informants). The remainder of this chapter will review these techniques. Examples will be drawn primarily from research on work and marital role stressors. Work and marital role stressors have been extensively studied by investigators using different types of self-report and observation measures; therefore, comparisons can be made among diverse methods of measuring chronic role stressors.

Theoretical Framework for Studying Role Stressors

Social roles provide boundaries that can help researchers to "gain some conceptual control over the extensive array of potential chronic stressors" (Pearlin, 1989, p. 245). Social roles are comprised of sets of interpersonal relationships, activities, duties, and responsibilities that are relatively easy to identify, tend to be stable, and affect a great number of people in the general population. Moreover, because of the great investment people make in social roles, especially work and marital roles, any difficulties or threats to functioning in these roles can be very stressful (Eckenrode & Gore, 1990; Lepore, Palsane, & Evans, 1991; Pearlin, 1989).

At the core of most studies of role stressors and health lie two major assumptions. A primary one is that some objective features of social roles, such as excessive physical and cognitive demands, or interpersonal tensions, are stressors and, as such, are related to physical and psychological health outcomes. A second assumption is that psychological appraisals (e.g., perceived harm, threat, loss) of objective role stressors are an important link between external, environmental stressors and internal psychological (e.g., depression, anxiety) or biological responses (e.g., blood pressure, neuroendocrine or immune alteration). Several variations of the basic stressor–appraisal–health model also can be identified, such as the specification of modifying factors that can make people more or less vulnerable to the stressors (see Kessler, Price, & Wortman, 1985). Modifier variables, such as actual and perceived control over role stressors, emotional and instrumental social support, and personality or disposition of the role incumbent (e.g., type A, optimism, negative affect), are presumed to alter role incumbents' perceptions of role stressors or their ability to adapt successfully to role stressors (Cohen & Edwards, 1989).

Two paragraphs cannot do justice to the rich theoretical models that have been developed in the study of role stressors and health. However, they serve to point out the important distinction between objective and subjective stressors that is apparent in most models of stress and health. The challenge to quantifying objective role stressors and the perceptions of those stressors is partly conceptual and partly technical. The conceptual challenge lies in identifying the objective and perceptual components of social roles that are relevant to health outcomes of interest. This task should be guided by past empirical work and plausible theoretical elaborations of why particular role stressors and perceptions should be linked to particular health outcomes (cf. Brown, 1989). An important technical challenge in quantifying the objective and subjective components of chronic role stressors lies in discriminating between the two components. More often than not, different researchers will use the same measure to represent either the objective role situation or a person's internal representation of the situation. As discussed below, the problem of confounding objective and subjective components of role stressors is largely due to overreliance on a single method of measurement, self-report questionnaires. Another important technical challenge is dating the onset and duration of objective role stressors and perceived stress, particularly in regard to the onset and progression of health problems (House, 1987; cf. Pearlin, 1989). This is an especially complex problem because both the sources of role stress and the health disorder can be chronic, and role stressors may have an ambiguous onset. These problems, and some potential ways to mitigate them, are discussed in the next section, in the context of particular methods of assessing sources of chronic stress in work and marital roles.

Methods of Assessing Work and Marital Role Stressors

Self-Report Questionnaires

The majority of researchers studying role stressors use self-report questionnaires that tap perceptions and attitudes about aspects of work or marital roles that may be stressful. In the work domain, there are many fairly reliable and multidimensional measures of role stressors, including the Work Environment Scale (WES; Moos, 1981), the Occupational Stress Inventory (OSI; Osipow & Spokane, 1987), the Occupational Stress Indicator (OSIn; Cooper, Sloan, & Williams, 1988), and the Job Content Questionnaire (JCQ; Karasek, 1985). The questionnaires are easy to administer and inexpensive to purchase, and have been used in many studies, thereby providing investigators with a standard for evaluating their own results. The WES differs from the other work stressor measures because it taps very global perceptions of the work environment as opposed to perceptions of particular aspects of the respondent's job. For example, an item on the JCQ is phrased "my job requires working very hard," whereas a similar item on the WES is phrased "employees work very hard." In the marital domain, too, there are many fairly reliable, inexpensive, and easy to administer multidimensional measures of role stressors, including the Family Environment Scale (FES; Moos & Moos, 1981); the Marital Situations Inventory (MSI; Smolen, Spiegel, Bakker-Rabdau, Bakker, & Martin, 1985); the Evaluating and Nurturing Relationship Issues, Communication and Hap-

piness scale (ENRICH; Fournier, Olson, & Druckman, 1983), and the Marital
Agendas Protocol (MAP; Notarius & Vanzetti, 1983). The FES has probably been
used more than any of the other measures in this class, but unlike the other measures, the FES does not focus on perceptions of particular individual-level marital
stressors. Instead, the FES measures the subject's global perceptions of the general
family environment.

Within the respective domains of work and marriage, the multidimensional self-
report measures assess many of the same types of stressors and often use similarly
worded items to measure them. The work stressor questionnaires typically include
items or scales related to role conflict, quantitative and qualitative demands or
overload, role ambiguity, lack of control or autonomy, lack of support or cohesion,
inadequate career opportunities, job insecurity, interpersonal conflicts, and responsibility for objects or people. The marital role stressor questionnaires typically
include items or scales related to problems in communication (e.g., spouse is
argumentative), verbal and physical abuse, lack of emotional closeness and affection, infidelity or sexual problems, excessive role demands or inequity in division of
labor (e.g., in child care or household chores), financial problems, difficulties with
relatives or friends of spouse, and inequitable distribution of decisionmaking power.
The multidimensional nature of the role stressor questionnaires is a strength: Investigators can assess a wide range of stressors within a role or they can assess just one
or two dimensions by using select subscales from a measure. The conceptual overlap in the scales suggests that they are somewhat interchangeable, with the exception of the WES and FES measures, as discussed above.

Role stressor questionnaires are often used to measure objective sources of
stress. However, as a self-report instrument, role stressor questionnaires depend on
a respondent's judgment, which inevitably is affected by biases related to his or her
emotional state, disposition, memory, or other personal characteristics. Because of
these subjective biases, self-report measures do not provide an entirely veridical
account of role stressors. Some self-reported information is, however, more veridical than other information, because it is less dependent on cognitive and emotional
processing (Frese & Zapf, 1988). Responses to items that require some evaluation
(e.g., "Did you work too many hours this week?") will be more strongly influenced
by subjective biases than items that request respondents simply to describe their
work conditions (e.g., "How many hours did you work this week?"). In response to
the evaluative item, a subject must remember how many hours he or she worked and
also perform some mental calculus to judge whether the number of hours worked
was excessive or not. Unfortunately, most of the items on role stressor questionnaires request subjects to make evaluations about their role stressors in addition to
describing them. Another factor that influences the objectivity of self-report measures is the degree of specificity of the items. When subjects are required to make
global perceptions of general role conditions, as in Moos's WES and FES, their
attitudes and affect will have a stronger influence on their judgments than when they
are asked to describe specific aspects of their role that are stressful.

Another problem with role stressor questionnaires is that none of them has been
shown to predict objective health outcomes in a consistent manner. There is evidence of positive associations between self-reported levels of exposure to role

stressors and role-related attitudes (e.g., dissatisfaction) and self-reported symptomatology (e.g., psychosomatic complaints, psychological distress). In general, people who report high levels of exposure to role stressors also tend to report negative attitudes about their role and high levels of symptoms. These findings, however, cannot be taken as evidence of the predictive validity of role stressors as measured by questionnaires. Associations between self-reported role stressors and negative role attitudes may be interesting in and of themselves, but negative role attitudes do not necessarily cause objective health problems (e.g., House, 1987). The associations between self-reported measures of role stressors and self-reported symptoms are problematic because they could be spurious, or driven by a "third variable" (e.g., neuroticism, negative affect), or because they are inflated from shared methods variance and item overlap (cf. Cohen, Kessler, & Gordon, Chapter 1, this volume; Ganster & Schaubroeck, 1991; Kasl, 1978; Spector, 1992). There are some relatively isolated examples (e.g., Evans & Carrere, 1991; House, Strecher, Metzner, & Robbins, 1986; Kiecolt-Glaser et al., 1993; Matthews, Cottington, Talbott, Kuller, & Siegel, 1987) of associations between self-report measures of role stressors and relatively objective health indicators (e.g., hospitalization and physician visits, blood pressure, smoking, cholesterol, neuroendocrine and immune functioning). However, these associations usually are found by investigators using different self-report role stressor measures, are seldom replicated when the same measure is used in different studies, and are often small in size. Therefore, it is difficult to find a single role stressor questionnaire that consistently predicts substantial variation in objective health outcomes.

Another major problem with role stressor questionnaires is that they seldom directly measure duration and frequency of exposure. Researchers often presume that because the stressful characteristics of some roles are stable, role incumbents' exposure to those stressors also will be stable. On the face of it, these appear to be reasonable assumptions, especially the presumed stability of role characteristics. For example, using nationally representative data, Karasek and colleagues found that workers' mean ratings of many self-reported work role stressors (e.g., job demands, job insecurity) did not change significantly over the years 1969, 1972, and 1977; and occupational-level analyses revealed very high stabilities (r's $> .90$) in the test–retest correlations of self-reported work role variables across the three survey years (see appendix in Karasek & Theorell, 1990). The level of stability in work role variables over time in these analyses is impressive because the data were based on ratings made by different respondents in different years. These data provide strong evidence that potentially stressful aspects of work roles are stable over time. The assumption regarding chronic exposure, however, is more tenuous. Duration of individuals' exposure to stressors depends on a number of factors that are seldom measured or controlled for in studies on role stressors. For example, people often change jobs and positions over time; thus there is great variability in duration of exposure to work role stressors across workers.

To test the effects of chronic role stressors thoroughly, variance in exposure must be systematically studied. One way to estimate exposure duration is to get information on how long a person has occupied a particular role. However, this is likely to be a very gross indicator of stressor duration because of changes in a person's role over time. For example, as a person becomes more senior in a work role, he or she is likely to acquire responsibilities that are not faced by a junior colleague. In marital roles, chronic stressors may not begin until there is a precipitous event, such as one spouse losing a job, contracting a terminal illness, or having a baby. The ideal approach for assessing duration is to take multiple measures over time. Prospective studies, or ones that use repeated measures, are particularly effective if initial measures of the outcome variable can be taken before a person is exposed to the stressors of interest. In this way, selection confounds can be tested and potentially ruled out when one is evaluating the association between exposure to role stressors and health outcomes.

Retrospective approaches, in which subjects recall exposures to past stressors, increase the risk of introducing measurement error due to forgetting. Forgetting, or falloff, in stressors occurs at a rate of approximately 5 percent per month on self-report checklists of stressful life events (Funch & Marshall, 1984). Falloff of items on role stressor questionnaires has not been examined, but it seems plausible that people will not forget chronic stressors as easily as they forget episodic ones (cf. Neilson, Brown, & Marmot, 1989). However, the risk of forgetting is always present in retrospective measures, especially if the stressor that is being remembered is not ongoing, or concurrent, with the interview. To avoid memory biases, it is best to measure concurrent exposure to stressors and to repeat the measures over time. Stressors that are enduring or recurrent can be considered chronic. The various methods of taking repeated measures are discussed in detail in the chapter on daily stressors (Eckenrode & Bolger, Chapter 4, this volume).

The subjective biases, inadequate predictive validity, and lack of attention to assessing duration in the more popular measures of role stressors make them a poor choice of measurement for investigators interested in studying objective, chronic stressors. However, it should be noted that role stressor questionnaires may be valuable for other purposes. For example, subjective evaluations of role stressors might be important criteria in intervention studies (e.g., marital counseling, job redesign). For those investigators who desire a more objective measure of chronic stressors, there are alternatives to self-report questionnaires, including observation and interviews.

Observation: Naturalistic, Conflict-Analogue, and Informant Sources

Observational approaches are a good alternative to self-report measures for investigators interested in obtaining objective estimates of stressors. Observational data have several potential strengths:

1. They are independent of the subject whose health is being predicted; thus they are relatively immune to the biases inherent in subjects' self-reports.
2. They can be collected from more than one observer and submitted to reliability analyses (e.g., inter-rater agreement).
3. They can be collected in real time; thus some of the biases caused by memory failures in retrospective reports are avoided.

Three general observational techniques will be discussed here: (1) naturalistic observation, (2) informant-based observation, and (3) laboratory analogues of conflict. In naturalistic observation, trained observers use standardized behavioral coding and rating scales to measure subjects' exposure to stressors as they participate in their normal role-related activities. In informant-based observation, people (e.g., co-workers, spouse) who have had a chance to observe subjects or to become familiar with their role estimate subjects' exposure to stressors on standardized scales. In laboratory analogues of conflict, trained observers use standardized behavioral coding and rating scales to measure stressful interaction patterns between people—usually spouses—discussing contentious issues (e.g., marital problems) in a laboratory setting.

Naturalistic observation is most feasible for investigators studying role stressors that have a high frequency and that can be readily observed. For example, work load can usually be estimated by trained observers because it involves aspects of a role that are often quite visible (e.g., amount, speed, intensity of role activities) and occur with some regularity. In studies that have assessed role stressors using both naturalistic observation and subjects' self-reports, there is some convergence between the measures, although it is modest. For instance, Kirmeyer and Dougherty (1988) found that trained observers' estimates of the frequency of work activities of police dispatchers (e.g., number of calls and transmissions) were moderately and positively correlated with the levels of work load estimated by the dispatchers themselves ($r = .35$). Frese (1985) has found convergence in two studies between trained observers' estimates of role stressors and subjects' self-report (r's ranging from .33 to .39) using composite measures of work role stressors that included items related to easily observed aspects of work (e.g., speed of work, physical intensity of work, time pressure) as well as more subtle aspects (e.g., role ambiguity, role conflict).

The convergence between role stressors measured by naturalistic observation and subjects' self-reports serves to validate both approaches to a degree. However, when put to the test of predicting health-related outcomes, naturalistic observation measures do not fare as well as subjects' self-report measures. For instance, in the Kirmeyer and Dougherty (1988) study, the self-report work load measure but not the observed work load measure was significantly and positively correlated with work-related anxiety. In the Frese (1985) studies, all self-report estimates of work stressors were significantly correlated with psychosomatic complaints (r's ranging from .31 to .40), whereas only two out of four of the observational measures were predictive of psychosomatic complaints (r's ranging from .18 to .19).

It is highly likely that shared methods variance explains the consistently higher association between two self-report measures than between observational and self-report measures. However, it is also possible that stressor data derived by naturalistic observation have weak predictive power because they can underestimate exposure to role stressors. Many role stressors may be hidden from an observer, particularly if the observer is an outsider who is familiar neither with the subject nor with his or her role. The presence of an outside observer also may cause subjects to alter their behaviors. The problem of reactivity to an observer is likely to occur in both work and marital situations. For instance, interpersonal conflict, which is a plausible stressor in work and marital roles, is a socially undesirable behavior that is likely to be suppressed in the presence of a researcher. Other sensitive or private aspects of work (e.g., negative work evaluations by a supervisor) and marriage (e.g., sexual difficulties) that are stressful frequently occur behind closed doors, where an observer has no access. Some role stressors may be underestimated because they are only partially visible to an outside observer. For example, work load may involve both manual and cognitive demands, but because cognitive demands are not easy to observe, they might not be counted in an observational measure of work load.

In the marital domain, investigators have used laboratory-analogue techniques to examine stressful interaction patterns that are difficult to observe in natural settings. An advantage of analogue techniques over naturalistic observation is that they do not require investigators to intrude into subjects' personal lives by entering their home. Instead, marital partners come into the investigator's laboratory, where they discuss marital problems or other disagreeable topics. The discussions are video-taped, audiotaped, or transcribed, and submitted to extensive coding (for detailed discussions of methods of coding marital behaviors, see the special issue of *Behavioral Assessment,* 1989; Jacob, 1987). There is some evidence that analogue measures can predict physiological outcomes that may be related to health. For instance, Kiecolt-Glaser and colleagues (1993) used a conflict-analogue technique to assess the effect of negative communication and interaction patterns (e.g., criticisms, interruptions, withdrawal) on immune functioning in newlywed couples. They found that subjects who exhibited a high number of negative behaviors during a discussion of marital problems had larger decrements in immune functioning over a 24-hour period and had larger and more prolonged increases in blood pressure during the discussion relative to subjects exhibiting few negative behaviors.

Although analogue techniques tend to be highly reliable and free of subjective biases, they have some limitations. One problem is that couples' behaviors tend to be constrained by the laboratory setting and task demands established by the investigator. For instance, couples who ordinarily would engage in physical aggression during conflict negotiations might suppress such behaviors in a laboratory (Margolin, Burman, & John, 1989). The analogue technique also does not provide information about the frequency or duration of actual problematic exchanges between couples, nor does it provide information about sources of marital role stress that are not interpersonal, such as financial difficulties. Therefore, analogue techniques reveal only a small number of the potential stressors that occur in a marriage.

In many respects, informant-based observation measures are better than naturalistic and laboratory-analogue measures of role stressors. Like the naturalistic and analogue observation data, informant-derived data are not tainted by the biases inherent in subjects' self-reports. However, because informants tend to be part of a subject's everyday life and immersed in the same environment, problems of reactivity are minimized and informants will have information about the subject's role stressors that are hidden from outside observers or not manifest in laboratory-analogue situations. Informants also may be aware of stressors that occur in relatively low frequency because of their familiarity with the subject's role. In addition, informants may be able to provide estimates of the frequency of exposure that would be difficult to obtain with naturalistic or analogue techniques.

In the work domain, Frese (1985) has compared informant methods to naturalistic observation and self-report methods. He found that informants' (co-workers) estimates of work role stressors were more strongly correlated with subjects' self-reports of work role stressors than were estimates derived from naturalistic observation. Informants' estimates of work role stressors also were more strongly correlated with subjects' psychosomatic complaints than were observers' estimates. However, of all the measures, the self-report measure was correlated most strongly with psychosomatic complaints. The best informants seem to be those who do the same job as the subject and who have had an opportunity to observe the subject at work. Supervisors can also be used as informants at work, but they often are lacking in knowledge of some role stressors. For example, Spector, Dwyer, and Jex (1988) found relatively strong correlations between secretaries' and supervisors' estimates of the secretaries' work load, as measured by number of hours worked ($r = .83$) and number of people worked for ($r = .65$), but found relatively little convergence between secretaries' and supervisors' ratings of secretaries' role ambiguity ($r = .08$) and role conflict ($r = .30$).

The lack of consistency in workers' and supervisors' ratings could be related to the degree of visibility of some stressors and not others. However, other factors also could be operating. For example, many times a supervisor is responsible either directly or indirectly for the stressors facing workers. Specifically, role ambiguity and role conflict could be a reflection of inadequate supervision. Thus, supervisors might be biased toward underestimating the amount of role ambiguity and conflict in their organization to project a more competent image of themselves. Similar biases might slip into co-workers ratings of stressors that might reflect negatively on them—especially stressors relating to interpersonal conflicts with co-workers. In addition, co-workers' knowledge about a subject's attitudes about work could bias how they rate the subject's exposure to stressors.

Similarly, when an informant is a spouse, it is possible that many of the items regarding marital-role stressors could relate to the informant's behavior in the marriage. Some of the behaviors that the informant engages in may not reflect positively on him or her. In addition, an informant may not want to admit that he or she is in a bad marriage. Informants who want to convey a positive image of themselves or their marriage might distort their reports about their partner's marital role stressors. One way around this dilemma is to have children serve as informants, if they are available. However, children may have their own biases and may also be a source of parents' marital problems. Not surprisingly, the most reactive marital measures are those that relate to embarrassing, sensitive, or highly personal topics (e.g., sexual dysfunction or violence). Arias and Beach (1987) found that men and women with high needs for social approval tend to report lower rates of physical aggression against their spouse than men and women with relatively low needs for social approval. However, they found no association between social desirability scores and reports of being the victim of a spouse's aggression (for discussions on the problem of differential reporting from husbands and wives, see Jacobson & Moore, 1981). When it is necessary to ask an informant sensitive questions, it is advisable to measure social desirability as a potential covariate, gather convergent sources of data, or use disguise techniques (e.g., bogus pipeline).

In addition to the substantive issues discussed above, many practical factors must be considered when choosing an observational method. Limited resources (e.g., personnel, time, money) can be a major constraint for many investigators. In comparison with informant-based approaches, naturalistic and analogue techniques require a great deal of resources and are difficult to implement. Much time and labor must be expended to train researchers in naturalistic and analogue observation techniques. In addition, when naturalistic observations are used, very long or repeated observations may be necessary to sample a sufficient number of stressful encounters involving the observed subject. When laboratory analogue techniques are used, the investigator creates a provocative situation, so it generally does not take much time to observe whether subjects have stressful interaction styles. However, after generating a conflict situation between subjects, coding the qualities of the stressful interaction can be very labor intensive. In addition, specialized and sometimes expensive video or audio recording equipment may be required in an analogue study. In contrast to the naturalistic and laboratory-analogue approaches, informant approaches shift the burden of data collection to a third party (i.e., the informant), so relatively little time and labor is required from the investigator.

Access to the population of interest is another practical consideration. Naturalistic observation may be perceived by subjects, or the people who control access to subjects, as intrusive. Researchers might not be permitted to visit work settings if managers fear that observers will interfere with work. Or, if a particular business is run like a sweatshop, researchers might not be welcomed because they could expose unsafe or illegal practices. Among married couples, the thought of an outsider observing their private life could be unsettling and offensive. As mentioned above, laboratory-analogue approaches might be perceived by couples as less intrusive than naturalistic observation. However, enticing couples to come into a laboratory can be an arduous task. It seems that problems of access would not be as great with informant techniques as with naturalistic and laboratory-analogue techniques. For instance, investigators who recruit informants to measure work role stressors would not have to disturb work processes because informants could provide their data during breaks, or before or after their work shift. Investigators measuring marital role stressors by means of an informant would not have to get a couple to come into the lab, which is time consuming, nor would they have to observe the couple at home, which can be intrusive.

Interview

Interviews can be more informative and precise than self-report questionnaires because interviewers can use probes, feedback, props, and cues to facilitate respondents' recall. Interviewers also can help to maintain subjects' attention so that they provide detailed information about the quality of their role stressors, the duration of stressors, the surrounding circumstances, the time sequencing of stressor exposure and health outcomes, and so on. Ultimately, however, interviews are self-report measures and therefore are subject to many of the same criticisms that have been leveled against questionnaires. Thus, interviews may be preferred over self-report questionnaires because they provide richer and potentially more precise data, but they generally will not be preferred to observational methods when the goal is to measure objective role stressors. If, however, the particular role stressor of interest is not observable by researchers or informants, then interviews might be the best measurement approach.

One of the most developed interview techniques for measuring stressors is the Life Events and Difficulty Schedule (LEDS; Brown, 1989; Brown & Harris, 1978; Wethington, Brown, & Kessler, Chapter 3, this volume). The LEDS approach uses a semi-structured interview to elicit detailed descriptions of life events (e.g., circumstances surrounding events, how events interrelate, time course of events) in various life domains (e.g., work, family) that are likely to "produce strong emotion of any kind" (Brown, 1989, p. 21). The LEDS is also used to measure ongoing stressors, called long-term difficulties (Brown & Harris, 1978). Long-term difficulties are generally measured by applying contextual ratings to life stressors that last at least 4 weeks. In addition to information about episodic and chronic stressors, LEDS interviewers gather social and biographical data on respondents (e.g., employment status, number of children, quality of social relationships). Trained raters then evaluate the likely meaning of an event to an individual, or normative degree of threat, given the nature of the event and the individual circumstances surrounding it. These contextual ratings measure whether a particular event poses a short- or long-term threat and the degree of threat, ranging from marked to little or none. By using investigator-based contextual ratings rather than subject's ratings of the magnitude of an event's stressfulness, the LEDS approach avoids many of the confounds and biases inherent in methods that rely on the subjects' ratings of events (see Wethington et al., Chapter 3, this volume).

Neilson and colleagues (1989) recently reported on a retrospective study that used the LEDS approach to examine the relation between work-related stressors and myocardial infarction (MI) in civil servants. Work-related events and difficulties were measured in ten 1-year periods before the interview. There was some falloff in reporting of stressors in more distal years. Falloff was greater for events (4.8% per year) than for chronic difficulties (4.1% per year), which suggests that chronic stressors are easier to remember than more acute ones. Neilson et al. (1989) developed two work-related stressor measures from the LEDS interviews, a workload index and a work stress index. A worker got a positive score on the workload index if in at least 5 of the 10 years preceding the interview he worked an average of 51 hours or more per week per year or took 2 weeks or less vacation time per year. A worker got a positive score on the work stress index if in at least 5 of the 10 years preceding the interview he had either a work difficulty or a work event involving long-term threat. The years containing difficulties greatly outnumbered those with an event (62:38); thus the work stress index was comprised mainly of ongoing work stressors. Workers who were positive on the workload index were more likely to have a first MI than workers who were not positive on the index (odds ratio = 5.36); workers who were positive on the work stress index also were more likely to have a first MI than those who were not positive (odds ratio = 4.18).

Despite its positive attributes, many investigators shy away from the LEDS because it is very time-consuming to administer and code, especially in comparison with informant-based observations and self-report questionnaires. However, as Neilson et al. (1989) demonstrated, the LEDS can be used to focus on selective stressors, such as work stressors. By limiting measurement to select stressors, or domains, investigators can save time in administering and coding the LEDS interview. Of course, some investigators may still be wary of using the LEDS because of theoretical concerns. The most controversial feature of the LEDS is the investigator-based contextual ratings (Tennant, Bebbington, & Hurry, 1981). The contextual ratings are based on data related to social and background characteristics of subjects that very well could have independent effects on health outcomes, or could moderate the effects of the life stressors. Thus, the unique effects of the life stressors independent of the situational and personal characteristics of the subject experiencing the stressors cannot be adequately determined. Brown (1989) has responded to this latter criticism primarily by arguing that alternative approaches are as yet not feasible or no more desirable than the LEDS approach. However, the very recent Structured Event Probe and Narrative Rating (SEPRATE; Dohrenwend, Raphael, Schwartz, Stueve, & Skodol, 1993) approach might resolve some of the problems associated with the contextual ratings of the LEDS.

Like the LEDS, SEPRATE uses a semi-structured interview to get in-depth narratives about different life stressors; however, unlike the LEDS, interviewers do not gather data on the stressful characteristics of the situation and the personal dispositions of the respondents. After the SEPRATE interview, trained judges rate the event on several dimensions (e.g., change, severity, fatefulness) based solely on the description of the stressful event or situation and how they believe the "average person" would respond to the stressful situation. Although the SEPRATE approach may avoid some of the problems of the LEDS contextual ratings, it is not clear how useful it will be for studying chronic stressors. The SEPRATE approach has focused mainly on assessing stressors that involve changes and disruptions in respondents' lives, rather than stable stressors, such as role-related demands and interpersonal conflicts. The LEDS approach, on the other hand, has been developed to look explicitly at both eventful and chronic stressors. In addition, there is evidence that the LEDS interview procedures improve long-term recall of events when compared with more traditional questionnaire or checklist approaches, irrespective of whether one uses the contextual rating procedure (Brown, 1989).

Conclusions and Future Directions

There is clearly no single method that adequately measures all chronic stressors. Observational methods are limited because they can be intrusive, causing people to alter their behaviors, or they do not allow investigators to capture the full range of stressors affecting a subject. Certain observational methods, such as naturalistic observations and laboratory analogues, are generally not feasible for investigators with limited resources. Observational methods are also of little use to investigators interested in measuring the subjective experience of chronic stressors. Self-report questionnaires can be used to measure stressors that are not easy to observe and to assess subjects' perceptions of stressors. However, responses to self-report questionnaires are often biased, particularly if the items focus on sensitive personal problems, involve long-term recall, or require subjects' to make judgments that can be influenced by their affective state or disposition.

Issues of feasibility have to be determined by individual investigators, but some guidance has been offered above. Often the easiest form of measurement is self-report questionnaires, but with a little more effort informants could be recruited together somewhat more objective data. Sometimes practical constraints, such as limited resources, can be overcome if two or more investigators are willing to combine their resources in a collaborative effort. Other problems, such as reactivity to sensitive questions and obtrusive observers, or biases related to memory and personal characteristics of the respondent, can be dealt with only by finding ways to adjust for the error that these problems introduce into measures of chronic stressors. The biasing effects of some variables can be controlled for with statistical techniques. For example, if need for social approval is suspected of distorting subjects responses on a questionnaire, it can be measured by use of a social desirability scale and then entered as a covariate in the analysis of the links between the stressor and health.

Another way to control for measurement error is to use a multi-method or multi-measure strategy of data collection (Cook & Campbell, 1979). The greatest advantage of using multiple methods for assessing chronic stressors is that it allows an investigator to develop an aggregate measure of chronic stressors that is devoid of measurement error. Consider, for example, the study by Frese (1985) discussed earlier in this chapter. Frese used three methods of data collection: self-report questionnaires, co-workers' observations, and naturalistic observations by trained personnel. Each of these methods has some imperfection, or error, but each also provides some unique and veridical information about workers' exposure to stressors. Frese chose to compare the effects of work stressors measured by the three different methods rather than to develop an aggregate measure of work stressors using data from the different methods. However, he very well could have taken advantage of the multiple methods by creating a latent variable. The three different methods used in his study were each designed to measure a single underlying variable: chronic work stressors. If a common factor underlies the three different measures, then it can be extracted by using confirmatory factor analysis. The resulting factor would be a latent variable (Loehlin, 1987), which would represent an error-free estimate of chronic work stressors. By reducing measurement error, investigators are often able to increase the predictive power of variables.

This chapter has focused mainly on measurement of discrete stressors, or stressors in a particular social role. However, some investigators might be interested in examining the health consequences of experiencing diverse chronic stressors. For example, Norris and Uhl (1993) have developed a questionnaire to assess chronic stressors in seven domains: marital, parental, filial, financial, occupational, ecological, and physical. They found that exposure to these chronic stressors seemed to mediate the long-term effects of an acute disaster (hurricane) stressor on psychological distress. In examining the association between diverse chronic stressors and psychological distress, Norris and Uhl did not form an aggregate measure of chronic stressors by summing subjects' scores across the 7 domains. Instead, they examined the independent effects of the diverse chronic stressors. This seems to be the most rational approach, even though investigators might be tempted to create a summary chronic stressor score. I recommend against aggregating scores from diverse chronic stressors for theoretical and analytical reasons. This recommendation is in contrast to the one made above for creating an aggregate score to measure a particular chronic stressor that has been measured by different methods. It is also in contrast to the usual practice of developing cumulative stress scores in life events research.

In chronic stressor research, in contrast to life events research, there is no major theoretical rationale for aggregating scores from diverse chronic stressors. In life events research, the theoretical precedent for summing scores from diverse acute life stressors is that life stressors all require a certain degree of change, or adjustment, in a person that can be estimated in terms of life change units (Cohen, Kessler, & Gordon, Chapter 1, this volume). A person's cumulative stress score is the sum of the weights of his or her life events, or total life change. A higher cumulative stress score reflects greater life change and, theoretically, will put a person at increased risk for health problems. This theoretical rationale does not transfer easily to the case of chronic stressors, partly because chronic stressors are not readily conceived of in terms of change. Indeed, many chronic stressors represent a lack of change: continuously oppressive work conditions, ongoing conflicts in the family, persistent noise, crowding, and air pollution, and so on. From an analytical perspective, it may be quite difficult to aggregate diverse chronic stressor scores. To the extent that chronic stressors from various life domains do not reflect an underlying construct, they will not form a latent variable (however, see Krause, 1990). Finally, analyzing an aggregate index of diverse chronic stressors would obscure underlying interactive and dynamic interrelations that may exist between different chronic stressors. For instance, chronic work stressors may exacerbate the health effects of chronic marital stressors or cause chronic marital stressors (Eckenrode & Gore, 1990). The way to observe such effects is to examine the unique and conjoint effects of chronic work and marital stressors on health outcomes and the interrelation between work and marital stressors.

In addition to dealing with problems of practicality and measurement error, future investigators of chronic stressors will have to attend more to issues of quantifying duration of exposure to stressors, as well as duration of appraised stress (cf. Baum, O'Keefe, & Davidson, 1990). The measurement of duration is inextricably linked to design issues. It seems that most chronic stressors will need to be assessed using prospective, repeated measures designs. Because of the fallibility of memory, duration of stressors measured in self-report questionnaires or by informants' reports are not very reliable for distal stressors or stressors that have been resolved prior to the data collection period. Therefore, when self-report questionnaires are used, an investigator should use multiple measurement periods with a short interval (e.g., 1 to several months) between measurements. The same recommendation holds for investigators using interviews. In comparison with self-report questionnaires, interviews may provide a more precise method for dating stressors, but interviews are not immune to memory biases. Finally, when naturalistic observation or laboratory-analogue techniques are used, duration of stressors can be determined only by extremely long observation periods or repeated observations over time.

Prospective, repeated-measures studies of chronic stressors not only will help to quantify the duration and frequency of exposure to stressors, but also will improve our ability to assess the plausibility of a causal relation between chronic stressor and health. The literature is currently overburdened with cross-sectional, retrospective, correlational studies that reveal little to nothing about the causal links between role stressors and health. However, prospective studies, especially when combined with multiple methods of measurement, will be quite costly. Therefore, this ideal approach will seldom be realized in stress research. Instead, most investigators will probably continue to rely on cross-sectional research designs and self-report questionnaires to measure stressors. One way to improve on this common research approach is to use objective health outcomes. This will reduce some alternative explanations about observed associations between the self-reported stressors and the outcome, such as a shared-methods variance explanation. It would also be useful, to the extent possible, to include some objective indicators of the stressors of interest. For instance, in a study of marital stressors, investigators could attempt to develop an index of financial stressors by gathering relatively objective information about employment, income, and expenses, in addition to subjects' self-reports of financial stressors in the marriage. Alternatively, an investigator who does not have the resources to collect observational data might conduct a limited number of observations to attempt to validate whatever self-report measure is being used in a study. In addition, the cross-sectional questionnaire study can be improved by measuring and statistically adjusting for variables such as negative affect, social desirability, and neuroticism, that might cause a spurious relation between the stressors and the outcomes. Of course, covariance procedures are somewhat limited, because all relevant covariates can never be identified. Nevertheless, covariance procedures, and each of the procedures just described, will be helpful toward improving the measurement of chronic stressors and the evaluation of their effects on health.

References

Arias, I., & Beach, S.R.H. (1987). Validity of self-reports of marital violence. *Journal of Family Violence, 2,* 139–149.

Avison, W. R., & Turner, R. J. (1988). Stressful life events and depressive symptoms: Disaggregating the effects of acute stressors and chronic strains. *American Journal of Health and Social Behavior, 29,* 253–264.

Baum, A., O'Keefe, M. K., & Davidson, L. M. (1990). Acute stressors and chronic response: The case of traumatic stress. *Journal of Applied Social Psychology, 20,* 1643–1654.

Brown, G. W. (1989). Life events and measurement. In G. W. Brown & T. O. Harris (Eds.), *Life events and illness,* (pp. 3–45). New York: Guilford Press.

Brown, G. W., & Harris, T. W. (1978). *The social origins of depression: A study of psychiatric disorders in women.* New York: Free Press.

Cannon, W. B. (1929). *Bodily changes in pain, hunger, fear, and rage.* New York: D. Appleton.

Cohen, S. (1980). Aftereffects of stress on human performance and social behavior: A review of research and theory. *Psychological Bulletin, 88,* 82–108.

Cohen, S., & Edwards, J. R. (1989). Personality characteristics as moderators of the relationship between stress and disorder. In R.W.J. Neufeld (Ed.), *Advances in the investigation of psychological stress* (pp. 235–283). New York: Wiley.

Cohen, S., Evans, G. W., Stokols, D., & Krantz, D. S. (1986). *Behavior, health, and environmental stress.* New York: Plenum Press.

Cook, T. D., & Campbell, D. T. (1979). *Quasi-experimentation: Design and analysis issues for field settings.* Chicago: Rand McNally.

Cooper, C. L., Sloan, S. J., & Williams, S. (1988). *Occupational Stress Indicator data supplement.* Windsor, UK: NFER-Nelson.

Dohrenwend, B. P., Raphael, K. G., Schwartz, S., Stueve, A., & Skodol, A. (1993). The Structured Event Probe and Narrative Rating method (SEPRATE) for measuring stressful life events. In L. Goldberger & S. Breznitz (Eds.), *Handbook of stress: Theoretical and clinical aspects* (2nd edition, pp. 200–261). New York: The Free Press.

Dooley, D., & Catalano, R. (Eds.). (1988). Psychological effects of unemployment. *Journal of Social Issues, 44*(4), 241–256.

Eckenrode, J. & Gore, S. (1990). Stress and coping at the boundary of work and family. In J. Eckenrode & S. Gore (Eds.), *Stress between work and family* (pp. 1–15). New York: Plenum.

Evans, G. W., & Carrere, S. (1991). Traffic congestion, perceived control, and psychophysiological stress among urban bus drivers. *Journal of Applied Psychology, 76,* 658–663.

Evans, G. W., & Cohen, S. (1987). Environmental stress. In D. Stokols & I. Altman (Eds.), *Handbook of environmental psychology* (pp. 571–610). New York: Wiley.

Fournier, D. G., Olson, D. H., & Druckman, J. H. (1983). Assessing marital and premarital relationships: The PREPARE–ENRICH Inventories. In E. E. Filsinger (Ed.), *Marriage and family assessment: A sourcebook for family therapy* (pp. 229–250). Beverly Hills, CA: Sage.

Frese, M. (1985). Stress at work and psychosomatic complaints: A causal interpretation. *Journal of Applied Psychology, 70,* 314–328.

Frese, M., & Zapf, D. (1988). Methodological issues in the study of work stress: Objective vs subjective measurement of work stress and the question of longitudinal studies. In C. L. Cooper & R. Payne (Eds.), *Causes, coping and consequences of stress at work* (pp. 375–412). New York: Wiley.

Funch, D. P., & Marshall, J. R. (1984). Measuring life events: Factors affecting fall-off in the reporting of life events. *Journal of Health and Social Behavior, 25,* 453–464.

Ganster, D. C., & Schaubroeck, J. (1991). Work stress and employee health. *Journal of Management, 17,* 235–271.

Herbert, T., & Cohen, S. (1993). Stress and immunity in humans: A meta-analytic review. *Psychosomatic Medicine, 55,* 364–379.

House, J. S. (1987). Chronic stress and chronic disease in life and work: Conceptual and methodological issues. *Work & Stress, 1,* 129–134.

House, J. S., Strecher, V., Metzner, H. L., & Robbins, C. (1986). Occupational stress and health among men and women in the Tecumseh Community Health Study. *Journal of Health and Social Behavior, 27,* 62–77.

Jacob, T. (Ed.) (1987). *Family interaction and psychopathology: Theories, methods, and findings.* New York: Plenum.

Jacobson, N. S., & Moore, D. (1981). Spouses as observers of the events in their relationship. *Journal of Consulting and Clinical Psychology, 49,* 269–277.

Jenkins, C. D. (1988). Epidemiology of cardiovascular diseases. *Journal of Consulting and Clinical Psychology, 56,* 324–332.

Karasek, R. (1985). *Job Content Questionnaire and user's guide.* Lowell, MA: University of Massachusetts.

Karasek, R., & Theorell, T. (1990). *Healthy work: Stress, productivity, and the reconstruction of working life.* New York: Basic Books.

Kasl, S. (1978). Epidemiological contributions to the study of work stress. In C. L. Cooper & R. Payne (Eds.), *Stress at work* (pp. 3–48). Chichester, UK: Wiley.

Kessler, R. C., Price, R. H., & Wortman, C. B. (1985). Social factors in psychopathology: Stress, social support, and coping processes. In M. R. Rosenzweig & L. W. Porter (Eds.), *Annual review of psychology* (Vol. 36). Palo Alto, CA: Annual Reviews.

Kiecolt-Glaser, J. K., Malarkey, W. B., Chee, M., Newton, T., Cacioppo, J. T., Mao, H., & Glaser, R. (1993). Negative behavior during marital conflict is associated with immunological down-regulation. *Psychosomatic Medicine, 55,* 395–409.

Kirmeyer, S. L., & Dougherty, T. W. (1988). Work load, tension, and coping: Moderating effects of supervisor support. *Personnel Psychology, 41,* 125–139.

Krantz, D. S., Contrada, R. J., Hill, D. R., & Friedler, E. (1988). Environmental stress and biobehavioral antecedents of coronary heart disease. *Journal of Consulting and Clinical Psychology, 56,* 333–341.

Krause, N. (1990). Stress measurement. *Stress Medicine, 6,* 201–208.

Lepore, S. J., & Evans, G. W. (in press). Coping with multiple stressors in the environment. In M. Zeidner & N. S. Endler (Eds.), *Handbook of coping: Theory, research, and applications.* New York: Wiley.

Lepore, S. J., Evans, G. W., & Schneider, M. (1991). Dynamic role of social support in the link between chronic stress and psychological distress. *Journal of Personality and Social Psychology, 61,* 899–909.

Lepore, S. J., Evans, G. W., & Schneider, M. (1992). Role of control and social support in explaining the stress of hassles and crowding. *Environment and Behavior, 24,* 795–811.

Lepore, S. J., Palsane, M. N., & Evans, G. W. (1991). Daily hassles and chronic strains: A hierarchy of stressors? *Social Science & Medicine, 33,* 1029–1036.

Loehlin, J. C. (1987). *Latent variable models.* Hillsdale, NJ: Lawrence Erlbaum Associates.

Margolin, G., Burman, B., John, R. S. (1989). Home observations of married couples reenacting naturalistic conflicts. *Behavioral Assessment, 11,* 101–118.

Matthews, K., Cottington, E., Talbott, E., Kuller, & Siegel, J. (1987). Stressful work conditions and diastolic blood pressure among blue collar factory workers. *American Journal of Epidemiology, 126,* 217–224.

McGonagle, K. A., & Kessler, R. C. (1990). Chronic stress, acute stress, and depressive symptoms. *American Journal of Community Psychology, 18,* 681–706.

Moos, R. H. (1981). *Work Environment Scale manual.* Palo Alto, CA: Consulting Psychologists Press.

Moos, R. H., & Moos, B. S. (1981). *Family Environment Scale manual.* Palo Alto, CA: Consulting Psychologists Press.

Neilson, E., Brown, G. W., & Marmot, M. (1989). Myocardial infarction. In G. W. Brown & T. O. Harris (Eds.), *Life events and illness* (pp. 313–342). New York: Guilford Press.

Norris, F. H., & Uhl, G. A. (1993). Chronic stress as a mediator of acute stress: The case of hurricane Hugo. *Journal of Applied Social Psychology, 23,* 1263–1284.

Notarius, C. I., & Vanzetti, N. A. (1983). The Marital Agendas Protocol. In E. E. Filsinger (Ed.), *Marriage and family assessment: A sourcebook for family therapy* (pp. 209–227). Beverly Hills, CA: Sage.

Osipow, S. H., & Spokane, A. (1987). *Occupational stress inventory: Manual research version.* Odessa, FL: Psychological Assessment Resources.

Pearlin, L. I. (1989). The sociological study of stress. *Journal of Health and Social Behavior, 30,* 241–256.

Seligman, M.E.P. (1975). *Helplessness: On depression, development, and death.* San Francisco: W. H. Freeman.

Selye, H. (1956). *The stress of life.* New York: McGraw-Hill.

Smolen, R. C., Spiegel, D. A., Bakker-Rabdau, M. K., Bakker, C. B., & Martin, C. (1985). A situational analysis of the relationship between spouse-specific assertiveness and marital adjustment. *Journal of Psychopathology and Behavioral Assessment, 7,* 397–410.

Spector, P. E. (1992). A consideration of the validity and meaning of self-report measures of job conditions. In C. L. Cooper & I. T. Robertson (Eds.), *International Review of Industrial and Organizational Psychology* (Vol. 7, pp. 123–151). New York: Wiley.

Spector, P. E., Dwyer, D., & Jex, S. (1988). Relations of job stressors to affective, health, and performance outcomes: A comparison of multiple data sources. *Journal of Applied Psychology, 73,* 11–19.

Tennant, C., Bebbington, P., & Hurry, J. (1981). The role of life events in depressive illness: Is there a substantial causal relation? *Psychological Medicine, 11,* 379–389.

Veroff, J., Douvan, E., & Kukla, R. A. (1981). *The inner American: A self-portrait from 1957–1976.* New York: Basic Books.

PART III

The Psychological Perspective

The psychological perspective on stress places emphasis on the organism's perception and evaluation of the potential harm posed by stimuli (stressors or events). The perception of threat arises when the demands imposed upon an individual are perceived to exceed his or her felt ability to cope with those demands. This imbalance gives rise to labeling oneself as being stressed and to a concomitant negative emotional response. It is important to emphasize that psychological stress is defined not solely in terms of the stimulus condition or the response variables, but rather in terms of the transaction between the person and the environment. Psychological stress involves interpretation of the meaning of an event and the interpretation of the adequacy of coping resources. In short, the psychological perspective on stress assumes that stress arises totally out of persons' perceptions (whether accurate or inaccurate) of their relationship to their environment.

Part III addresses the measurement of appraisal (Chapter 6) and affect (Chapter 7), two concepts that have been closely linked to the psychological perspective. Measures of appraisal focus on persons' evaluations of events, of their own coping resources, and of their perceptions regarding whether they are experiencing "stress." Although central to the psychological approach, there has been little work on the development of adequate measures of appraisal, and thus Chapter 6 focuses on the difficult issues involved in appraisal measurement as much as it does on the limited number of instruments. In contrast, there is a voluminous literature on the measurement of emotion. Chapter 7 discusses various ways of thinking about emotions (e.g., negative vs. positive, activated vs. unactivated) and identifies different emotional states. It provides information on a range of measurement techniques and their strengths and weaknesses for addressing specific questions about the role of emotion in the disease process.

6

Measurement of Stress Appraisal

Scott M. Monroe and John M. Kelley

Preliminary Considerations

It is almost axiomatic that for an event or situation to be stressful it must be apprehended or perceived by the organism. From this perspective, environmental challenge is thought to be mediated largely via perceptual processes; in turn, these perceptual processes are hypothesized to be influential with respect to health and well-being. Most generally stated, then, stress emerges from the perceptual interface between world and person, the product of which is related to subsequent susceptibility to disorder. Additionally and more practically, the pivotal position of perception in stress theory helps explain individual differences in stress responses and outcomes. Why can two people, when confronted with very similar types of life events, differ dramatically in their emotional reactions and adaptations? For instance, two individuals who are laid off from the same jobs with the same company may perceive the situation quite differently. One could view the termination of employment as an opportunity for change and career enhancement; the other individual could view the termination as the "last straw" in a string of seemingly unbearable adversities. Although there may be many additional factors influencing such differences in viewpoints adopted, there is a strong and enduring theoretical interest in this pivotal role of perception and its prominence in affecting the subsequent stress process.

Much of current stress theory revolves around the cognitive process of appraisal, a more refined concept than perception that pertains to the evaluation of stressors and their dimensions (Lazarus, 1966; Lazarus & Folkman, 1984). According to Lazarus and Folkman (1984): "Cognitive appraisal can be most readily understood as the process of categorizing an encounter, and its various facets, with respect to its significance for well-being . . . it is largely evaluative, focused on meaning or significance, and takes place continuously during waking life" (p. 31). Building on this notion, psychological stress "is a particular relationship between the person and the environment that is appraised by the person as taxing or exceeding his or her resources and endangering his or her well-being" (p. 19) (see also Chapter 1).

Measuring an individual's appraisal of stress appears to be a very straightforward means of testing cognitive models of stress and their implications for health and well-being. In this context it is rather bewildering that there are relatively few measures currently available. In a recent discussion of this topic, Lazarus (1991) noted:

> The central concept of my theoretical analysis of psychological stress is *appraisal*. So in our research, Folkman and I (see Lazarus & Folkman, 1987, for a review) made some limited progress in its measurement, but the procedures we used were much too primitive to survive long in that form. I hope they will stimulate further refinements. A few scattered researchers have taken up the challenge of advancing the measurement of appraisal for use in predictive research. (p. 446).

Thus, although there are different methods available for measuring appraisal (which we shall review in greater detail subsequently), the options are relatively limited and the empirical findings sparse compared to many of the other stress factors reviewed in this volume. One must therefore ask: Why is there such a paucity of measures for what many regard as the conceptual core of the stress process?

One reason might be that, given the primary theoretical position accorded appraisal in stress research, many of the issues so often debated within general stress theory are intensified at this level of analysis. When the theoretical emphasis is placed on appraisal, it brings into sharper focus many of the difficult problems that are dealt with more diffusely in the general literature on life stress. In a sense, the perennial definitional difficulties plaguing the stress concept have been supplanted and condensed within the notion of appraisal. The surface clarity of the concept gives way to diverse, sometimes competing, underlying ideas. These underlying ideas reflect different nuances of the concepts involved with appraisal, as well as the other factors that, although related to appraisal, might best be differentiated from it. If these theoretical issues are not adequately specified, then the measures based upon such theory will be compromised as reflections of the underlying constructs. The paucity of existing measures of appraisal, then, can be viewed as a symbol of our lack of adequate understanding of these matters at a more fundamental level. It is not surprising, then, that existing measurement approaches are limited, and that specific measurement practices differ, depending upon which particular facet of the conceptual space related to appraisal is adopted by a particular investigator.

Viewed from a more optimistic perspective, however, research on appraisal processes holds real promise for shedding light on the very problems that have historically beleaguered stress research. By bringing these concerns into sharper focus, the concept of appraisal and its measurement may clarify theoretical concerns and empirical relations with diverse outcomes believed to be stress related. Thus, the use of existing procedures to develop more informative empirical bases, along with the refinement of existing methods, could be of considerable benefit in elucidating fundamental issues in stress research. As a consequence of the relative lack of accepted measurement practices, however, the contents of the present chapter differ from others in this volume. Although we review existing measures of appraisal in detail, the relative lack of such instruments and available data requires that we first address several issues of relevance for clarifying the concepts involved and for facilitating future measurement refinement and development. Our goal is to provide a general context for thinking about measuring stress appraisal, as well as an evaluation of the existing options for its measurement.

History of Measurement Approach

The idea that a person's perception of worldly events is at least as important as, if not more so than, the events themselves has venerable origins. It can be found in the writings of the ancient Greeks through more modern times, most often with reference to the origins and influences of emotions in general (see Lazarus, 1991; Solomon, 1980). The more delimited concept of *appraisal,* in terms of the individual's evaluation of the meaning of encounters with the environment, has a more recent heritage, and is more specifically linked to the development of stress theory. As traced by Lazarus and Folkman (1984), many early writers on psychological stress "made use of the concept of appraisal, although mostly in an unsystematic, informal way or by implication" (p. 25). Grinker and Spiegel, in their 1945 classic *Men Under Stress,* employed the term "appraisal" in its modern, restrictive usage (i.e., "appraisal of the situation requires mental activity involving judgment, discrimination, and choice of activity, based largely on past experience"; p. 122). According to Lazarus and Folkman (1984), Arnold (1960) was the first to develop a systematic treatment of the concept of appraisal. Other writers, too, incorporated features of appraisal in their thinking, either explicitly (e.g., Erdelyi, 1974; Mandler, 1975), or implicitly (e.g., Janis & Mann, 1977). In general, strands of the concept of appraisal and related themes are woven throughout recorded history on the topic of emotion in general and, more recently, of stress in particular.

Such widespread interest in processes related to and bearing upon appraisal reflects the evolving importance attached to the general notion. But the diversity of allusions and approaches to the concept also indicates a lack of consensus on the ideas involved. By far the most systematic and well-delineated treatment of the topic has been undertaken by Richard Lazarus and colleagues (Lazarus, 1966, 1991, 1993; Lazarus & Folkman, 1984; Lazarus & Launier, 1977). Having already shown an early interest in psychological stress and human performance (e.g., Lazarus, Deese, & Osler, 1952), Lazarus and colleagues began a series of laboratory studies in which they attempted to alter the manner in which subjects construed challenging situations (e.g., stressful motion pictures). These experimental manipulations resulted in different patterns of psychophysiological responses. These initial studies were derived from ego-defense theory; the conceptual focus of the work subsequently shifted from ego-defense theory to "the broader concept of cognitive appraisal and reappraisal by means of which people constantly evaluated the realities of their experience and also protected themselves from threat" (Lazarus, 1991, p. 141).

These studies were quite informative about the possible role of appraisal in the stress process; yet the effects of stress were only *assumed* to operate via appraisal (i.e., there was no actual measurement of appraisal). These studies were complemented by retrospective reports about what subjects thought and felt under stress, as well as by research in which subjects were selected on the basis of personality or cognitive style to evaluate differences in thinking and feeling under stress (see Folkman & Lazarus, 1984; pp. 39–40). As this line of study progressed, specific definitions pertaining to the appraisal process were delineated. Of particular importance is the distinction between primary appraisal and secondary appraisal. Although they are not regarded as strictly separate processes, these two kinds of appraisal are important "to differentiate between discrete and complementary sources of knowledge on which evaluation of the personal significance of an encounter rests" (p. 133; Lazarus, 1991).

Primary appraisal involves the evaluation of the environmental situation with regard to the person's well-being. Three types of primary appraisal are posited: (1) irrelevant, (2) benign–positive, and (3) stressful; it is stressful appraisals that are the focus of present interest. Within this latter category, there are at least three types of stress appraisals: (a) harm/loss, (b) threat, and (c) challenge. *Harm/loss* pertains to situations in which some damage or loss to the individual has occurred; *threat* involves anticipated or possible future damage or losses; and *challenge* refers to situations that present the possibility for growth or gain (see Lazarus & Folkman, 1984, for further details and distinctions). *Secondary appraisal* pertains to the capabilities of the individual for dealing with the situation: "It is a complex evaluative process that takes into account which coping options are available, the likelihood that a given coping option will accomplish what it is supposed to, and the likelihood that one can apply a particular strategy or set of strategies effectively" (Lazarus & Folkman, 1984; p. 35). In summary, there is an iterative process over time in which secondary appraisals feed back upon primary appraisals; it is this interaction over time that constitutes the appraisal process and modulates the degree of stress experienced (Lazarus & Folkman, 1984).

It is noteworthy that the concept of coping, too, is interwoven with the appraisal process. Whereas at any one point in time secondary appraisal is influenced by the person's perceived ability to cope with the event, over time the actual coping activities and their efficacy play into the appraisal process in an important way. There have been many attempts to measure various dimensions of coping activity (e.g., Billings & Moos, 1981; Folkman & Lazarus, 1980; Moos & Schaefer, 1993; Pearlin & Schooler, 1978). Yet theoretically it is important to keep separate the concept of coping from that of appraisal (see Chapter 1). Furthermore, although many related measurement issues are involved with the assessment of coping, additional considerations pertain in such assessment that go well beyond the measurement of primary and secondary appraisal (e.g., see Moos & Schaefer, 1993; Stone, Greenberg, Kennedy-Moore, & Newman, 1991). Thus, while we acknowledge the complex interplay over time between appraisal and coping, we confine our focus to measures of appraisal.

Derogatis and Coons (1993) have pointed out that transactional theories of stress have been slow to lead to new measurement procedures "in large part because of the inherent difficulties of measurement in a constantly changing system" (p. 202). More recently, however, a limited number of investigators have adopted the basic appraisal framework and undertaken empirical studies. Despite the diversity of appraisal dimensions studied (e.g., control, undesirability, anticipation, and so on; see Schwartz & Stone, in press), the manifold operational schemes implemented (e.g., up to 49 different appraisal scoring schemes have been examined; see Neale, Hooley, Jandorf, & Stone, 1987), and the different formats developed (e.g., single-item questions vs. more broad-based questionnaires; see Cohen, Kamarck, & Mermelstein, 1983; Peacock & Wong, 1990), the collective findings from these studies are encouraging. (The specific measures are reviewed in greater detail in the section below, "Types of Questions the Measures of Appraisal Can Answer.") Overall, although the ideas are long-lived and fertile, the history of the measurement approach is more recent, limited, and diverse in related concepts and procedures. Most recently, translation of the ideas into empirical form provides a sound basis for further research and development.

Rationale for the Measurement Approach

As discussed by Cohen, Kessler, and Gordon (Chapter 1, this volume), the psychological model of stress emphasizes the organism's view of the situation: "The perception of threat arises when the demands imposed upon an individual are perceived to exceed his or her felt ability to cope with those demands." This perception, or appraisal, is accorded a central position between the stressor and the organismic responses that eventuate in disease. (Although it is commonly assumed that appraisal results in emotional responses that have biological concomitants and consequences of relevance for vulnerability, it is important to note that there are other possible pathways of effects that could also result in compromise of the organism's adaptive capabilities. For example, other behaviors such as coping or avoidance may promote vulnerability independently of negative affective states.) The key issue for present purposes is that the environmental situation needs to be evaluated psychologically as *meaningful* to the organism.

There are several virtues to the psychological model, which we detail in the next section. When we turn to translating such ideas into specific operational procedures and testable hypotheses, however, a number of important considerations arise. In a sense, the flexibility and fertility of appraisal theory currently outstrip the existing technologies for measuring the phenomena of hypothetical relevance. We view these issues as central to the development of adequate procedures for measuring the appraisal of stress, and consequently devote attention to their discussion in the second section below.

Rationale and Support for the Concept of Appraisal

There is an elegance to measuring stress in terms of a person's appraisal. *If* such processes are pivotal in translating environmental adversity into psychological and biological sequelae of relevance to health and well-being, then one can abstract the essence of the encounter and study its implications in a relatively straightforward manner. Directly targeting this core feature avoids the messiness of independently measuring the myriad characteristics of the environment that lend themselves to perception on the one hand, as well as the many organismic dimensions that influence perception on the other hand. In theory, one can encapsulate such diverse influences into the final determining product, or "final common pathway," of appraisal, and then move forward in investigating the likelihood of subsequent dysfunction (Depue, Monroe, & Shackman, 1979; Lazarus & Folkman, 1984).

The dynamic and transactional features of appraisal, too, represent an important

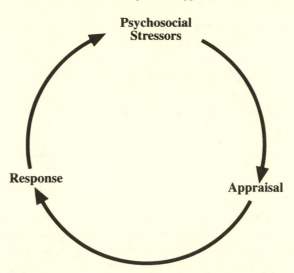

Figure 6.1. Schematic diagram of the transactional model of stress and the appraisal process.

step forward in addressing the complex evolution of stress as a product of repeated interactions between person and world. Rather than viewing stress in a simplistic and static manner, such a formulation provides theoretical flexibility to accommodate more complicated patterns of transaction between person and environment that may represent more accurate depictions of the stress process. Figure 6.1 provides a simplified schematic diagram of the transactional model of stress and the appraisal process. The organism is confronted with psychosocial stressors, appraises the stressors, and responds. In accord with the transactional model of the stress process (Lazarus & Folkman, 1984; Lazarus & Launier, 1977), these responses in turn influence the stressors, which in turn are reappraised, and so forth through an evolving cycle of appraisals, responses, and alterations in the stressors. This model of the stress process helps capture the evolving and dynamic nature of the appraisal process over time.

As we have noted already, recent research has yielded promising leads on the feasibility and utility of measuring appraisal. For example, Tomaka, Blascovitch, Kelsey, and Leitten (1993) reported that measures of appraisal derived from Lazarus and Folkman's appraisal model of stress were predictive of both physiological and behavioral effects. Schwartz and Stone (in press) found that measures of specific appraisal dimensions were important in understanding the relations between type of stressor and choice of coping strategy. Cohen, Tyrrell, and Smith (1993) reported that heightened levels of perceived stress prospectively predicted important aspects of the development of the common cold.

Finally, the idea of appraisal and its importance finds related support in the literature on cognitive therapy and the treatment of diverse emotional problems (Beck, Rush, Shaw, & Emery, 1979). The theory of cognitive therapy is predicated on the premise that particular cognitive processes contribute to maladaptive emotional and behavioral responses. The interventions derived from this theory are designed to change underlying cognitive structures or schemas and their more consciously accessible thinking patterns and products. Thus, the rationale for appraisal processes in the stress model is consonant with findings from different, although conceptually related and important, research literatures and traditions.

Concerns Related to the Rationale Underlying the Concept of Appraisal

There are four issues that, with some discussion, may help to provide a conceptual and methodological context for appraisal measurement. First, it is not very clear how appraisal is to be differentiated from other aspects of cognition at any one point in time. Second, and related, the transactional nature of appraisal in the stress process intensifies the concerns with differentiating appraisal from other components of cognition over time. Third, the costs and benefits of employing appraisal as a singular, short-hand "final common pathway" representation of stress require some discussion. Finally, there are additional issues pertaining to the general approaches adopted for assessing appraisal. For example, should one directly assess subjective judgments about appraisal by the respondent, or should one utilize more objective approximations of the person's appraisal as developed by the investigator?

Distinguishing Appraisal From Other Cognitive Processes. Despite the common usage of the appraisal idea in discussions of stress theory and research, existing definitions and references to the concept ultimately are unclear with regard to what is *not* potentially subsumed within the construct. It is difficult to draw the boundaries between appraisal and the other aspects of the person's phenomenology or ongoing awareness. For example, it is not apparent how appraisal differs from anxious rumination, worry, fear, catastrophizing thoughts, distress intolerance, or other aspects of awareness. Indeed, appraisal can be loosely linked not only with the vast cognitive content of consciousness, but also with processes outside of conscious awareness (e.g., defense mechanisms, nondefensive attentional processes; see Lazarus & Folkman, 1984). Thus, although one can state in general terms what appraisal is (i.e., evaluating the meaning of environmental events and circumstances), it is not so easy to delineate where this construct departs from other constructs of the mind. Much of the individual's consciousness that is either irrelevant to the stress process or is an artifact of other cognitive processes from which appraisal must be differentiated can be included within the general concept of appraisal.

There are two general considerations. First, it may be that appraisal is simply a "noisy" construct as currently formulated and measured. This suggests that although there is a strand of "truth" in the position, there is random error that dilutes the definition and measure. Although Lazarus and colleagues have distinguished three appraisal components (i.e., harm/loss, threat, challenge), it is not readily apparent how these hypothetical dimensions are distinguished from other attributes of the individual's cognitive response. In addition, other investigators have proposed other appraisal dimensions in recent research. Figure 6.2 provides a schematic example of the problem.[1] In response to any particular stressor, there are a variety of cognitive

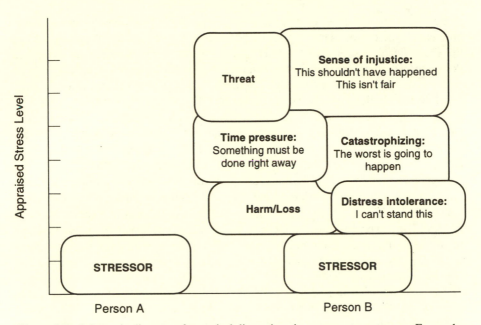

Figure 6.2. Schematic diagram of appraisal dimensions in response to a stressor. For explanation, see text.

themes that may develop. For example, Person A in Figure 6.2 is essentially without appraisal and represents an extreme example for comparative purposes; he or she simply responds to the "objective" event. Person B, on the other hand, displays many different cognitive themes in response to the stressor. Some of these are traditional themes proposed for appraisal (e.g., threat, harm/loss), whereas others are not (e.g., time pressure; sense of injustice; distress intolerance). There are few guidelines on how such diverse components of cognitive activity are to be grouped within the appraisal construct and across different appraisal dimensions, or distinguished as independent facets of cognitive activity.

Alternatively, appraisal may be confounded with other cognitive processes as currently formulated and measured. In this case, there is a strand of "truth" within the general notion, but there is also a good deal of error that could be systematically associated with a range of dependent variables typically investigated in stress research. The appraisal process is influenced by many factors, which in turn may be predictive of health and/or well-being. For example, personality factors, psychopathology, and mood state of the person all likely affect appraisal and the reporting of stress (Aldwin, Levenson, Spiro, & Bosse, 1989; Cohen, Towbes, & Flocco, 1988; Lazarus & Folkman, 1984), and such factors may independently influence susceptibility to disorder.

Perhaps most importantly, there is the thorny issue of differentiating appraisal from psychological distress (i.e., a maladaptive or extreme psychological state). Although it appears feasible that one can remove the "distress" from appraisal by studying appraisal processes within nondistressed populations, the converse proposition seems questionable. Can one take the "stress" out of "distress"; can one study appraisal if the person is actively distressed? There may be means of distinguishing the theoretical constructs of "appraisal" and "distress" at an empirical level, but their interplay over time makes the exercise a most difficult one. And although there may be interesting arguments addressing such concerns about causal priorities and possible tautologies between the concepts (see Lazarus, 1991), the basic concern remains that these critical debates have not entered into the operational and empirical arena in a satisfactory or systematic manner (cf. Cohen & Williamson, 1988).

It should be acknowledged that considerable and valuable thought has been devoted to clarifying the construct of appraisal and its relations to other constructs of relevance (e.g., Lazarus, 1991; Lazarus & Folkman, 1984), and useful procedures have been developed for distinguishing appraisal from psychological distress (e.g., Cohen & Williamson, 1988). Our point is that despite the fact that some clarity in the ideas involved has been achieved, nontrivial issues concerning discriminant validity remain. Overall, our present goal is to point to important additional avenues of thinking necessary for further measurement development.

The Transactional Nature of Appraisal. Since appraisal is viewed as a process that evolves over time, some of the problems already discussed become more important to resolve. This is particularly so for the manner in which appraisal is to be distinguished from other variables that might influence both appraisal and the dependent variables under study. Of foremost importance is the possibility that for some people appraisal is distorted or biased. Probably the most dramatic example involves understanding how the appraisal process proceeds for depressed individuals. This is illustrated in a general manner in Figure 6.3 (which is an extension of the general transactional model portrayed in Figure 6.1). From this diagram it can be seen that psychopathology can influence all stages of the transactional model. For instance, it is well known that depressives have a biased perception of their worlds, selves, and

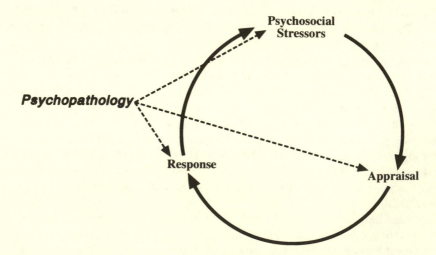

Figure 6.3. Schematic diagram of how psychopathology affects the appraisal process.

future (Beck, 1967). It is quite likely that the depressed person's appraisal of psychosocial stressors will be skewed, owing to his or her cognitive-affective condition. Furthermore, the coping capabilities of depressed individuals is compromised, thereby making responses to stressors more difficult and less effective. Finally, depressed persons often bring about the very stressors they are then forced to endure. For example, irritability, poor concentration, lack of energy—along with other hallmarks of depression—all make serious psychosocial stressors in employment or interpersonal relations more likely to occur (e.g., poor work performance, troubles with supervisors, difficulties with spouse, separation, etc.). Within this scenario, one must question whether or not appraisal is a functionally important component of the process or a by-product of more basic and influential processes. Stated differently, under these conditions appraisal may be "overdetermined" by other, more parsimonious explanations, rendering the discriminant validity suspect. Overall, it appears important to question the validity of appraisal as measured from persons in the presence of manifest distress.

Once we entertain the idea that appraisal can be influenced by a variety of factors, some of which suggest important differential implications for refining the construct with regard to later health and well-being, additional considerations can come into play. The implications of a "final common pathway," as currently formulated, require examination. Under certain conditions the final common pathway may be confounded (e.g., by preexisting distress) on the one hand, or may encapsulate "too much" (e.g., coping and stress resolution) on the other hand. This viewpoint opens up the appraisal process for finer analysis in terms of its determinants and constituent components. We next address some of the related considerations involved.

Appraisal as a Final Common Pathway. Appraisal may be influenced by a variety of factors. Personality variables, psychopathology, cognitive styles, beliefs, values, and current mood state are but some of the more relevant considerations (see Lazarus & Folkman, 1984). Although this viewpoint of the appraisal process provides flexibility in describing the evolution of appraisal, it also presents some definitional concerns. If there are a variety of influences of appraisal, one must question how much of the effect of stress is attributable to the appraisal per se versus the different factors that give rise to the appraisal.

A problem with a singular focus on appraisal in the stress process is that one cannot tease apart the causal role of the determinants of appraisal from the causal role of appraisal alone. In theory, we have indicated that appraisal tends to be viewed as the "final common pathway" in which diverse influences are integrated and the meaning of the encounter is synthesized (Depue et al., 1979; Lazarus & Folkman, 1984). Major environmental adversity can be minimized by appraisal, with professed healthful effects; minor annoyances can be amplified by appraisal, with ostensible insalubrious consequences. Without an equal emphasis on assessing the antecedents of appraisal, though, one is forced into an agnostic position with regard to the causal priority of appraisal. Without additional information, there are at least four competing models of stress effects that cannot be distinguished.

This issue is most readily demonstrated by simplifying an individual's total

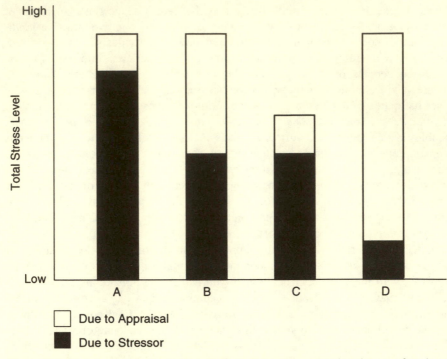

Figure 6.4. Four hypothetical cases in which final total stress level varies as a function of two major components—appraisal and stressor.

stress level into two major components: appraisal and stressor.[2] Figure 6.4 portrays four hypothetical cases in which final total stress level varies as a function of these two components.[3] Person A has a high total stress level, which is based on a relatively severe stressor yet relatively low appraised threat. Person B has a total stress level equal to that of person A, along with a relatively less threatening stressor but greater degree of appraised threat. Person C has a lower total stress level, yet has a stressor comparable to that of person B and an appraisal comparable to that of person A. Finally, person D has a high total stress level equivalent to those of persons A and B, yet has the least threatening stressor of the four hypothetical cases. A model based on total stress level alone would suggest that the vulnerability of person A would equal those of persons B and D, all of whom would be more susceptible than Person C. A model based on the stressor alone would posit person A as most vulnerable, person B and person C as equal in vulnerability, and person D as lowest in vulnerability. A model based on appraisal alone would predict person D as most vulnerable, person B as next most vulnerable, with person A and person C as least vulnerable.[4]

Yet the manner in which stressor and appraisal interact may be very relevant, and in turn can modify these simple models in several alternative ways (Monroe & Simons, 1991). As Lazarus and Folkman (1984) suggest, "[P]eople are normally constrained in what they perceive and appraise by what is actually the case, although their cognitions are not perfectly correlated with objective reality" (p. 48). An "environmental constraint" model would predict that given equal levels of environmental demands, appraisal processes will be influential. Put more concretely in terms of Figure 6.4, person A would be most vulnerable, person B next most vulnerable—importantly distinguished from person C (who has a comparable stressor, but less threatening appraisal—and finally person D least vulnerable. This would suggest that individual differences in appraisal become most relevant *once individual differences in the stressor are taken into account*. Alternatively, an "appraisal dominance" model would posit person D as most vulnerable, with person B next most vulnerable, and person A next in vulnerability—as distinguished from person C who, despite appraisal comparable to person A, has a less severe stressor. Such an ordering with respect to vulnerability in this model is dictated by the premise that individual differences in the magnitude of the stressor become most important once individual differences in appraisal are taken into account.[5]

Without an understanding of the antecedents of appraisal, the investigator cannot make very penetrating statements about the etiologic mechanisms or role of appraisal in the stress process (Lazarus & Folkman, 1984). It is quite likely that the importance of appraisal may be masked in much of current research, owing to the lack of control of other factors of relevance (e.g., stressor magnitude). Ultimately, the question concerns how much appraisal is a crucial link in the causal pathway, a mediator nested within dimensions or levels of environmental adversity, a consequence of other factors that are more etiologically essential, or a tautological finding with regard to the outcome variable (i.e., stress predicting distress via distress). Put differently, appraisal may be a reflection of underlying forces that are more directly responsible for incurring susceptibility rather than a determinant of such susceptibility itself. As such, appraisal may be a "marker" rather than a "maker" of vulnerability. To the extent that appraisal is a "marker" of vulnerability, the inclusion of other elements of the model is required to determine the essential processes involved with producing vulnerability. To the extent that appraisal is a "maker" of vulnerability, the inclusion of other elements in the model is required to substantiate such conjecture. Thus, by incorporating measures of the antecedents and components that contribute to appraisal, one can better test such competing views of the role of appraisal in the stress process.

General Approaches to Assessing Appraisal. The final issue we raise with regard to the measurement of appraisal pertains to the manner in which the information is to be obtained and categorized. The most popular approach is for the individual to provide the information in its final form. Since it is only the person who has the necessary scope and awareness of his or her motives, commitments, concerns, and so on, that impart meaning to the situation, he or she is the best arbiter of appraisal. Such "inside-out" or subjective measures of appraisal are based on the premise that it is only from within the perceived world of the person that the true meaning of the event can be understood.

An assumption of this "inside-out" viewpoint for measuring appraisal is that people's "on-line" reporting of appraisal adequately captures the cognitive dimensions of crucial relevance for understanding the stress–disorder link. If this assumption is valid, this represents a very important strength of the approach (e.g., increased accuracy and sensitivity to the specific circumstances of the particular person involved). Yet there are potential problems with this assumption that must be confronted directly. As we have already indicated, there is a vagueness about what is and is not included within the domain of appraisal. One must ask whether or not naive subjects can provide such judgments in a reliable and accurate manner. For example, after major adverse life events, it is common for the individual to go through phases of oversensitization, denial, intrusive recollections, and so on (e.g., see Horowitz, Bonanno, & Holen, 1993). In a sense, cognitive and emotional lability is to be expected in appraising and adapting to major events. A rejected lover may vacillate between feelings of pain and abandonment, pride and self-righteous indignation. Overall, would appraisal judgments measured at one point in time be replicable at other points in time under circumstances of varying mood and processing of the situation?

An alternative approach is to approximate the likely appraisal of the person, but to avoid the potential subjective contaminants that go with the subject-based approach. This "outside-in" or investigator-based method employs trained raters to judge the likely impact of an event or situation on the average person existing in comparable biographical circumstances (Brown & Harris, 1978; see also Chapter 3). With detailed assessment of the individual and his or her life situation, the approach may become sensitive to the major dimensions of relevance contributing to appraisal, yet avoid the methodological pitfalls of the subject-based approach. The cost is in effort and potential sensitivity. The interview must be highly detailed and time consuming, and the raters must have sophisticated training and understanding of the measurement system involved. Even given such effort, the question remains as to whether or not such a procedure can adequately capture the idiographic circumstances and nuances of meaning that are theoretically the essential ingredients of appraisal.

Types of Questions Measures of Appraisal Can Answer

Our discussion thus far has attempted to outline additional considerations for measuring appraisal and to present possibilities for developing a framework for interpreting appraisal influences. Such information, combined with the available appraisal measures, provides a reasonable base for addressing several types of questions.

Although true of essentially all measures covered in this volume, it bears repeating that measures of appraisal confine the investigator to study primarily the prediction of disorder. This is in contrast to the study of the mechanisms and dynamics involved in susceptibility and etiology of disorder. Consequently, the data from such work permit only estimates of the *association* between appraisal and disorder, with the causal and etiologic importance of the construct remaining a major question. In the terminology of our discussion above, current measures allow only the level of analysis of possible "marker" of disorder vulnerability as a function of appraised stress as opposed to possible "maker" of such vulnerability. Yet this is still an important research agenda, given the fragmented and limited existing information on measured appraisal, stress, and disorder.

To the extent that other antecedents of the stress model can be incorporated into the research design, the stronger the research footing is for developing a more pene-trating understanding of the processes involved and the role of appraisal. The central theoretical position of appraisal within the stress model suggests that essentially all of the other measurement strategies addressed in this volume are of relevance. Although most studies cannot incorporate such a wide array of assessments, it is relatively easy to incorporate an index of appraisal within stress studies. With the limitations of these measures both psychometrically and conceptually kept in mind, an empirical base can be developed to help guide future measurement approaches and inquiry. Thus, if possible, it would be useful for investigators of diverse inter-ests to attempt to measure appraisal within their studies of stress and its particular effects.

In regard to research that more specifically targets appraisal, we suggest some priorities on additional measures to be included. In particular, it would be useful to include both an index of more objective, or consensually defined, environmental adversity (e.g., see Chapters 2 to 5) along with measures of subjective appraisal. At our current level of knowledge, it is important to begin to establish the degree to which any particular situation faced by the individual imposes meaning on the encounter versus the degree to which the individual imputes meaning to the situa-tion. Only through independent measurement of appraisal and its antecedents can one avoid the redundancy in measurement that has characterized much of stress research (see also Kasl, 1978; Monroe & Steiner, 1986). Additionally, it is most important to collect measures of initial emotional status of the subject (e.g., see Chapters 7 & 8). Information from such assessments would be helpful for placing the importance of appraisal in the wider context that we have outlined above. Other effects of stress (e.g., biological indices; Chapters 9 & 10) might be incorporated as per the specific interests of the particular investigator.

Choosing an Appropriate Measure

Types of Measures Available

There are four general classes of appraisal measures that can be discerned from the existing literature. The first class involves primarily single-item questions designed to measure appraisals of specific stressors. The majority of these have been devel-oped ad hoc for a particular study. The second class represents more broadly based self-report scales. These instruments attempt to assess the degree to which the specific or collective situations in one's life are appraised as stressful, and typically are composed of multiple-item inventories. The final two classes of measures are derived from life events research: self-report life event scales and interview-based measures of life stress. These life events instruments may have a direct, or indirect, bearing on the measurement of appraisal (see Chapters 2 & 3 for more detailed reviews of the specific instruments). Thus, although these approaches were not developed initially to assess appraisal per se, they can provide information of relevance for the construct.

Ad Hoc Single-Item Measures

This class of measures probably constitutes the largest body of evidence on measures specifically targeting appraisal. As noted by Moos and Schaefer (1993), "In general, appraisals have been measured with one-item indexes that assess individuals' immediate reactions (threat, challenge, harm, or benefit) to the situation and the extent to which it can be changed or must be accepted" (p. 242). These measures tend to be situation specific, and thereby assume that the immediate stimulus context is very influential in appraisal (as opposed to, but not excluding, influences by more enduring characteristics of the person). Such measures typically require rather immediate assessments of appraisal following exposure to the situation. Measures of this type are commonly used in laboratory experiments or in studies of daily events using diary methods. (Although some investigators have developed and used measures of emotions that correspond to particular appraisal dimensions [e.g., Folkman & Lazarus, 1985, 1988], we do not include these under appraisal measures because they inherently confound appraisal with coping and emotions; see Peacock & Wong, 1990.)

Examples of such appraisal measures can be found in the recent literature. Some of these measures distinguish between primary and secondary appraisals, whereas others do not. For example, in the study by Schwartz and Stone (in press), subjects were requested to answer eight appraisal questions related to the "most bothersome event or issue of the day" (covering control, undesirability, change, anticipation, meaningfulness, chronicity, prior experience, and stressfulness); the investigators found appraisals to predict different coping strategies (e.g., seeking social support, catharsis, and relaxation). Neale et al. (1987) report a series of studies involving the assessment of daily experiences, including studies of appraisal of daily events using up to 49 different appraisal classification schemes. The findings suggested the utility of the appraisal measures for predicting mood. Tomaka et al. (1993) conducted a study in which primary and secondary appraisal of stress was examined in three laboratory experiments. The stressor was counting backwards from a predesignated large number (e.g., 1,528) by intervals of 7. These investigators assessed primary appraisal by asking subjects, "How stressful do you expect the upcoming task to be?" (Experiment 1) and "How threatening do you expect the upcoming task to be?" (Experiments 2 and 3). They assessed secondary appraisal by asking subjects, "How able are you to cope with this task?" (for all three experiments). (Both questions used a 7-point Likert-type scale.) Cognitive appraisal of stress was operationalized in final form as a ratio of primary appraisal to secondary appraisal, and overall was found to predict a variety of relevant outcomes in conjunction with different task demands. (See also Dobson & Neufeld, 1981; Folkman & Lazarus, 1985, 1986; Folkman, Lazarus, Dunkel-Schetter, Delongis, & Gruen, 1986; for a more complete review, see Peacock & Wong, 1990.) Overall, these examples represent measures that are at least conceptually appropriate for assessing contemporaneous perceptions of appraisal. They possess face validity with regard to appraisal, are temporally synchronized with the beginning phases of the process, and have predicted relevant outcomes.

Although such measures provide useful preliminary information about appraisal with regard to the stress process, there are basic psychometric limitations to be borne in mind. For example, little is known regarding the reliability of such indices or the degree to which they are free from various sources of measurement bias (e.g., demand characteristics, current mood state). Little is known, too, about measurement characteristics for diverse populations (e.g., sex differences, socioeconomic status [SES] differences, and age differences). Single-item measures, too, are likely to possess high measurement error. As summary measures of the appraisal process, they represent only one "standard" point in time; therefore, they do not capture the dynamic, changing features of the construct. It is doubtful that they would be of use in retrospective studies, owing to the likelihood of increasing memory distortion or difficulties as one moves away in time from the referent period. That there are predictive findings from such measures suggests some degree of validity, particularly in situations where the results have been replicated (e.g., Tomaka et al., 1993). Such work provides a useful base of reference for research on appraisal and for the development of more comprehensive and psychometrically sound measures of appraisal.

Multiple-Item Scales

In contrast to situation-specific approaches to assessing appraised stress, other investigators have been interested in developing more elaborate instruments that target appraisal or dimensions of appraisal. There are two types of measures that have been developed.[6] First are measures of a specific stressor and the associated appraisals. Second is the type of measure that targets appraisal as a response to the cumulative total of life stressors facing the individual. Representing a more global or general, appraisal, this latter category incorporates the theoretically diverse influences of appraisal, ranging from the environmental context through cognitive styles, personality, and mood state. Both of these types of measures are, in theory, well suited to a variety of research contexts (e.g., laboratory studies through field research).

The Stress Appraisal Measure (SAM) was developed specifically to assess three dimensions of primary appraisal (threat, challenge, and centrality) and three dimensions of secondary appraisal (controllable-by-self, controllable-by-others, controllable-by-anyone) for a specific anticipated stressor (Peacock & Wong, 1990). Preliminary findings suggest that the appraisal dimensions possess relatively strong psychometric qualities (e.g., good internal consistency; the different dimensions are only moderately intercorrelated) and concurrent validity (e.g., expected associations with overall stressfulness and psychological symptoms). As Peacock and Wong (1990) suggest, "[T]here is need for further psychometric data, especially those obtained in differing contexts and with a broader range of respondents" (p. 235). Thus, the utility of the measure for different subject populations also requires further study (e.g., different age levels; gender differences; SES differences; cultural differences). Additional work, too, is required to determine the discriminant validity of the measure with respect to mood states and psychopathology. Yet the SAM appears to be one of the few instruments developed in a systematic manner designed specifically to assess dimensions of primary and secondary appraisal and which explicitly attempts to distinguish coping processes from appraisal processes (see also Fish, 1986; Gall & Evans, 1987). Other versions of the scale have been adapted by the SAM creators for use with ongoing and past events. Finally, there may be limitations inherent in scales purporting to assess appraisal of specific stressors, for some evidence suggests that people may misattribute feelings of stress from one source to another more salient one (see Cohen & Williamson, 1988).

The only empirically established index of which we are aware that falls into the category of general appraisal instruments is the Perceived Stress Scale (Cohen et al., 1983). The PSS, too, was developed based on Lazarus's concept of appraisal (Lazarus, 1966; Lazarus & Folkman, 1984): "The PSS measures the degree to which situations in one's life are appraised as stressful" (Cohen et al., 1983; p. 385). The 14 items of the original scale "were designed to tap the degree to which respondents found their lives unpredictable, uncontrollable, and overloading" (p. 387), and was intended for use in community samples with at least a junior high school education (Cohen & Williamson, 1988). The early studies using the PSS found the measure to possess good psychometric qualities (e.g., adequate reliability and predicted associations with other indices of stress). More recently, extensive normative data on 2,387 respondents are available for not only the original 14-item version of the PSS, but also 10-item and 4-item versions, which provide a rich reference base for studying perceived stress across gender, SES, age groups, race, and other demographic characteristics (Cohen & Williamson, 1988). (This measure has also been translated into Spanish and Italian.) Although all three versions provide strong psychometric data and are related to relevant outcomes in expected ways, Cohen and Williamson (1988) note the relative superiority of, and therefore recommend, the 10-item version. Most interestingly and importantly, recent studies have demonstrated prospective associations between perceived stress as measured by the PSS and a variety of relevant outcomes, even when other relevant predictors are controlled for in the analyses (see Cohen & Williamson, 1988; Cohen et al., 1993). In particular, these investigators recently found that perceived stress (as measured by the PSS) prospectively predicted important aspects of the common cold, and did so *differently* from other indices of stress:

> Negative life events were associated with greater rates of clinical illness, and this association was primarily mediated by increased symptoms among infected persons. Perceived stress and negative affect were also related to clinical illness, but their associations with increased risk were primarily attributable to increased infection. These differences suggest that (a) the negative life events instrument measures something different than perceived stress and negative affect and (b) the constructs they tap have somewhat different consequences for the pathogenesis of infectious illness. (Cohen et al., 1993:138)

The major limitation of the PSS is also one of its virtues. As a general measure of perceived stress, the scale is influenced by a wide variety of factors. Of greatest concern is the likelihood of overlap between psychological symptoms and what is measured by the PSS. Cohen et al. (1983) note this caveat, reporting correlations of .76 and .65 between the PSS and a measure of depressive symptoms (CES-D; Radloff, 1977) for two samples. Yet these investigators also found differential prediction afforded by the PSS once other predictors, including measures of psychological symptoms, were statistically controlled (suggesting adequate discriminant validity) (Cohen & Williamson, 1988). However, the magnitude of correlations reported in these studies is very comparable to the average intercorrelation found between different self-report measures of depressive symptomatology (see Clark & Watson, 1991, for a comprehensive review). One could argue that with the availability of other or more extensive measures of psychological symptoms, the differential prediction afforded by the PSS could be attenuated. Thus, although the PSS has been found to predict various outcomes independently of measures of psychological symptoms (Cohen & Williamson, 1988; Cohen et al., 1983, 1993), there may be room for further conceptual and method refinement with regard to discriminant validity of the PSS and general measures of psychological symptomatology. Finally, as we have indicated in our discussion above, such a summary measure, when used alone, makes it difficult to disentangle the antecedents of appraisal and thereby to set the stage for understanding the role of appraisal conceptually vis-à-vis its antecedents.

Life Events Scales

A number of scales designed to assess recent life events make provisions for assessing the subject's perceptions about events that may have occurred in their lives (e.g., the Life Experiences Survey [Sarason, Johnson, & Siegel, 1978]; The Impact of Event Scale [Horowitz, Wilner, & Alvarez, 1979]). As such, these measures conceivably could serve as a general index of appraised stress (to the extent that the particular life event inventory covers the relevant domains of potential stressors), or as an index of specific appraised stress for any particular event that has been experienced. We could also include in this category the Hassles Scale, an instrument assessing relatively minor daily events (Kanner, Coyne, Schaefer, & Lazarus, 1981). This measure, developed from appraisal theory and transactional models of stress, was designed to measure the more proximal, immediate experiences of daily life. Although it could be argued that measures such as the Hassles Scale and its derivatives provide a more sensitive reflection of appraisal processes than traditional life event checklists, there are a number of additional method issues involved (e.g., see Dohrenwend, Dohrenwend, Dodson, & Shrout, 1984; Lazarus & DeLongis, 1985; Monroe, 1983). Since these measures of more minor daily events suffer from many of the same problems of the general major life event checklists, we do not treat them separately.

Problems with such approaches to the measurement of appraisal include the established difficulties with operationalization of the life stress by self-report methods (Brown, 1981; Dohrenwend, Link, Kern, Shrout, & Markowitz, 1987; McQuaid, Monroe, et al., 1992; see also Chapter 2). Additionally, research on differential weighting schemes for life events inventories indicates that such procedures tend not to enhance prediction (Derogatis & Coons, 1993; Lei & Skinner, 1980), and similar findings have been reported for daily events weighted by different appraisal dimensions (see Neale et al., 1987). Not all events are perceived as very stressful, and not all events necessarily lead to disorder. But many low-level events might equal a few major events that are perceived as extremely stressful with such simple operational schemes (Monroe & Simons, 1991). At this point, it seems prudent to question as well the degree to which subjects' appraisals are sufficiently standardized, calibrated, and reliable for assessing the individual events. Finally, once again there is the issue of distinguishing the appraisal from its antecedents. Thus, for both the total event score and the individual item score, these indices cannot be recommended as measures of appraisal.

Investigator-Based Approaches

In contrast to the self-report life event scales, interview-based methods that include definitional rules, guidelines, and operational definitions for events and difficulties, provide a better standardization basis for assessing life stress (see Chapter 3). The most well established of these measures is the Life Events and Difficulties Schedule (LEDS; Brown, 1989; Brown & Harris, 1978). Within the LEDS system, all events are rated based on extensive information about the circumstances surrounding the event and on the particular individual's biographical situation. Raters, however, are blind to the individual's subjective response. This provides what is termed "contextual ratings" (see Brown 1989; Brown & Harris, 1978). The manual provides approximately 5,000 case examples that describe features of the event (or difficulty) as well as brief relevant biographical context of the individual. These examples allow raters to "anchor" their rating of an event with standardized descriptors and ratings that are sensitive to features of the situation that are of importance for understanding the likely meaning of the event for the individual.

The relevance of this approach for present purposes is that it provides a relatively sensitive index of the likely meaning of the event for the average person in similar circumstances, yet avoids the majority of methodological concerns associated with contamination of mood state, personality, and so on with appraisal. Conceivably one could derive a summary index based on such an approach (e.g., the total of events rated to be at a particular level on the contextual scale). Although the definitional foundation for such ratings would be stronger than that for life event checklists, the index would likely lack sensitivity for the same reasons already discussed with regard to life event checklists (i.e., mixing diverse events that may dilute the final index; see Monroe & Simons, 1991). The approach would appear to be better suited to gauging the likely appraisal of single events (or difficulties).

The shortcomings of such an approach include the potential lack of sensitivity. It may be that the private understanding of the situation, coping capabilities, and other facets of the individual's phenomenological sphere are necessary for capturing the essence of appraisal. Additionally, the appraisal model does not incorporate the transactional issues involved (i.e., the estimates are based on the initial impact of the event, not on how the situation evolves). Finally, the standardization procedures (i.e., case exemplars and anchor ratings) have been developed primarily on females, who tended to be young and of lower social class; clearly further work with other populations is needed. Nonetheless, such measures might be of use and relevance, particularly given the method concerns inherent in other more subjective approaches. Such information provides one basis for beginning to account for individual differences in consensually defined adversity, and thereby to better understand the role of appraisal within such contexts.

Logistical Issues

There are very large differences in terms of logistical issues between the existing measures of stress appraisal. The ad hoc measures are clearly quite simple and convenient, but one must bear in mind the psychometric limitations. The SAM, on the other hand, represents a promising measure for assessing primary and secondary appraisal dimensions involving a specific stressor. The scale is composed of 37 items, which suggests that it does not require a great deal of time to administer. Of course, the scale can be used only for subjects who are likely to experience, or have already experienced, the specifically designated stressor (Peacock & Wong, 1990).

Alternatively, the PSS provides a very expedient measure that has garnered a good deal of supportive evidence. This is true of the original 14-item scale as well as the abbreviated versions, most especially the 10-item scale. As we have noted, however, the discriminant validity of the index must be of concern when one is investigating outcomes pertaining to emotional well-being and psychological status, or to other physical health outcomes where negative affect may be an important alternative predictor. Despite these limitations, this measure represents the best available index of general appraisal and can be incorporated relatively easily into a variety of research contexts.

In terms of life events measures, self-report indices vary in length and detail, but tend to be relatively expedient; subjects can complete them rather readily, and they require little investigator time. As we have discussed above, however, the utility of such measures as an index of appraisal is quite doubtful. With regard to interview-based measures, the index of long-term contextual threat derived from the LEDS requires extensive time from both the subject and the investigator. The interview can take between 1 and 3 hours, and the separate ratings by blind judges can require several additional hours (with both estimates dependent upon the time period of reporting adopted by the investigator and the degree of environmental adversity in the respondent's life). The training for both interviewer and blind judges is also rather extensive. Yet this measure provides potentially useful perspective on appraised stress, by obviating many of the methodological concerns that beset other indices (but again, does so with the potential disadvantage of losing sensitivity to the specific details and appraisals of the specific individual). Although not recommended for general studies of stress and its effects, measures of relevance for appraisal such as those afforded by the LEDS are necessary for developing a better understanding of the antecedents of appraisal, the contaminants and confounds associated with appraisal, and the ultimate role of appraisal in the stress process.

Future Directions

We have outlined what we would describe as the "first generation" of stress apprais-al measures. Although derived from theory, the different appraisal measures cur-rently available vary greatly with respect to the care taken in their development, their psychometric characteristics, the requirements placed on the subject, and the degree to which they are correlated or confounded with other components of the stress process. Since measurement is an essential component for construct explora-tion and validation (Loevinger, 1957), our knowledge of appraisal is not yet well informed by empirical evidence. To advance theory on stress in general, and on appraisal in particular, investigators need to devote more extensive research and devise additional approaches to the measurement of appraisal.

There are some promising leads with regard to measurement. In particular, with regard to general appraisal measures, the PSS has proven to be a useful instrument in a variety of contexts for predicting different stress-related outcomes (Cohen & Williamson, 1988; Cohen et al., 1983). It is possible that the PSS could be refined, or related measures developed, to enhance discriminant validity. For example, items might be dropped that are very highly correlated with psychological symptoms (or worded almost identically to such items in symptom scales), and additional items added that tap other dimensions of stress. Such "second generation" measures might be able to differentiate more precisely the *perceived* stress from the *manifest* dis-tress, and thereby refine the measurement of appraisal. With regard to measures of stressor-specific appraisals, there is a need to move beyond single-item ad hoc measures that vary from study to study. The SAM (Peacock & Wong, 1990) may be a promising development in this regard, but it requires more extensive research. Further work on the different dimensions of appraisal is necessary, targeting differ-ent types of psychological and physical outcomes, to clarify further which aspects of appraisal might be most relevant for particular types of problems. Finally, more complex and labor-intensive procedures may be useful to capture more precisely the dynamic interplay among emotions, defenses, and appraisal over time (see, e.g., Horowitz et al., 1993).

Clarification of some basic definitional and conceptual matters should comple-ment the development of better appraisal measures. Although there has been a great deal of work dedicated to these definitional and conceptual issues, questions remain and must be debated. As we have seen, Lazarus and colleagues have been relatively specific about the appraisal dimensions associated with stressors. More rigorous approaches to assessing these dimensions are clearly needed. In addition, other investigators have proposed other dimensions of potential relevance for appraisal. The nature of these dimensions and their relationship to the appraisal process requires elaboration. Overall, we have tried to point out some of the assumptions and ideas that have been woven into current thinking about appraisal, as well as the ways in which such concepts have been translated into operational form. We hope that through an enlightened examination of such issues, the generation of additional empirical information, and the translation of such concerns into refined measure-ment practices, a better understanding of appraisal in relation to the stress process will be forthcoming.

Notes

1. We thank Anne D. Simons for the idea and development of Figure 6.2.

2. Stress level represents the *combination* of the stressor (i.e., environmental input) and the appraisal of the stressor. Both the event and appraisal can vary, and the final stress level is a function of both components. The premise we adopt here is that people's stressors differ in many ways, some of which are likely to be extremely relevant for understanding individual differences in appraisal. It is only when we begin to take into account, in a systematic manner, the variation in people's stressors that we shall be able to understand better the manner in which such events *lend themselves* to appraisal processes. Lazarus and Folkman argue (1984): "In order to understand variations among individuals under comparable conditions, we must take into account the cognitive processes that intervene between the encounter and the reaction, and the factors that affect the nature of this mediation. If we do not consider these processes, we will be unable to understand human variation under comparable external conditions." (p. 23). In a parallel line of reasoning, we believe that if we do not actively take into account the environmental variation (i.e., what are "comparable external conditions"), we shall not be able to clarify the processes involved with appraisal. This does not mean that we are adopting a strictly positivist position with regard to the primacy of an "objective" environment. Rather, we are attempting to discriminate between degrees of normative, or consensually defined, environmental stressors. Thus, although we do not subscribe to the viewpoint that there is an "objective" environment or stressor (see Lazarus, 1991), we do subscribe to the viewpoint that environments and stressors often differ across individuals, and that such differences may be of great importance for developing an understanding of appraisal and undertaking an analysis of stress.

3. We emphasize that we have simplified the situation to illustrate our central point (i.e., that to achieve a better understanding of appraisal processes, one must have a better grasp of that which is being appraised). Two point, though, are worth noting. First, under normal conditions we would expect a closer correspondence between the magnitude of the stressor and the magnitude of the appraisal (i.e., the more severe the stressor, the more threatening the appraisal). This would result in a more evident functional relationship between the magnitude of the life event and the degree of appraisal. Second, we have restricted appraisal to a purely additive representation in the figure. However, appraisal could *reduce* the stress level by attenuating the magnitude of the stressor (i.e., by denial or intellectualization, thus minimizing the objective event). Thus, we have selected hypothetical cases that are likely to be somewhat atypical of the relationship between a life event and an appraisal. Nonetheless, these modifications help to illustrate our central point, and also the complexity of the issues involved, even at such a reduced level.

4. Given stress theory, it might seem that person D would be an unlikely candidate for "most vulnerable." Yet it will be useful to distinguish appraisals of specific stressors from styles of appraisal that cut across a variety of stressors. Thus, although person D may have the least severe stressor, the extreme degree of appraised stress may reflect an enduring disposition to evaluate a myriad of life situations and stressors as highly meaningful and threatening, thereby generating chronic negative affect and potential susceptibility.

5. We could easily argue for other alternative models of susceptibility, depending upon the particular assumptions made about appraisal, stressor, and their functional relationship. In the examples provided thus far, we have simply considered appraisal and stressor to be independent of each other. However, there is an asymmetry between the "environmental constraint" model and the "appraisal dominance" model with regard to the likely association between stressor and appraisal. Specifically, it is difficult to consider appraisal to be independent of what is to be appraised (i.e., the stressor). (Note that for the "environmental con-

straint" model, we believe it is reasonable to assume that the stressor arises independently of the appraisal, although counter arguments can be made; see Monroe & Simons, 1991; Simons, Angell, Monroe, & Thase, 1993.) This asymmetry suggests some interesting differences in the manner in which alternative models might be portrayed. Based on the assumption that appraisal at least in part follows from the nature of the stressor, we could posit that the ordering of the "appraisal dominance" model should require that person D be most vulnerable, with person B, C, and A following in order of increasing susceptibility. (The questionable comparison for the two "appraisal dominance" models is between persons A and C, who have roughly equal appraisal entering into their different stress levels.) Because person C appraises a less severe stressor equivalently to person A's appraisal of a more severe stressor, it could be argued that person C has a greater propensity to appraise events as threatening, and therefore is more likely to experience heightened levels of stress over time, given comparable stressors. (Alternatively, one could describe person A as minimizing stressors over time, thereby lessening cumulative vulnerability over time.) The validity of any particular assumptions and speculations aside, the general point is that one must take into account such variation in stressors in the face of variation in appraisal in order to understand the elements operative in the process that may lead to susceptibility to illness.

6. Other measures for assessing general appraisal processes associated with emotional experience have been developed for studies not specifically targeting stress appraisals (see Lazarus, 1991; Smith & Ellsworth, 1985, 1987). These approaches may be useful for clarifying dimensions of appraisal and their possible implications for the stress process, yet represent a different theoretical emphasis and therefore are not included in our review.

References

Aldwin, C. M., Levenson, M. R., Spiro, A. I., & Bossé, R. (1989). Does emotionality predict stress? Findings from the normative aging study. *Journal of Personality and Social Psychology, 56,* 618–624.

Arnold, M. B. (1960). *Emotion and personality* (Vols. 1 & 2). New York: Columbia University Press.

Beck, A. T. (1967). *Depression: Clinical, experimental, and theoretical aspects.* Philadelphia: University of Pennsylvania Press.

Beck, A. T., Rush, A. J., Shaw, B. F., & Emery, G. (1979). *Cognitive therapy of depression.* New York: Guilford.

Billings, A. G., & Moos, R. H. (1981). The role of coping responses and social resources in attenuating the stress of life events. *Journal of Behavioral Medicine, 4,* 139–157.

Brown, G. W. (1989). Life events and measurement. In G. W. Brown, & T. O. Harris (Ed.), *Life events and illness* (pp. 3–45). London: Guilford Press.

Brown, G. W. (1981). Life events, psychiatric disorder, and physical illness. *Journal of Psychosomatic Research, 25,* 461–473.

Brown, G. W., & Harris, T. O. (1978). *Social origins of depression: A study of psychiatric disorder in women.* New York: The Free Press.

Clark, L. A., & Watson, D. (1991). Tripartite model of anxiety and depression: Psychometric evidence and taxonomic implications. *Journal of Abnormal Psychology, 100,* 316–336.

Cohen, L. H., Towbes, L. C., & Flocco, R. (1988). Effects of induced mood on self-reported life events and perceived and received social support. *Journal of Personality and Social Psychology, 55,* 669–674.

Cohen, S., Kamarck, T., & Mermelstein, R. (1983). A global measure of perceived stress. *Journal of Health and Social Behavior, 24,* 385–396.

Cohen, S., Tyrrell, D.A.J., & Smith, A. P. (1993). Negative life events, perceived stress, negative affect, and susceptibility to the common cold. *Journal of Personality and Social Psychology, 64,* 131–140.

Cohen, S., & Williamson, G. M. (1988). Perceived stress in a probability sample of the United States. In S. Spacapan & S. Oskamp (Eds.), *The social psychology of health.* Newbury Park, CA: Sage.

Depue, R. A., Monroe, S. M., & Shackman, S. L. (1979). The psychobiology of human disease: Implications conceptualizing the depressive disorders. In R. A. Depue (Ed.), *The psychobiology of the depressive disorders: Implications for the effects of stress* (pp. 3–20). New York: Academic Press.

Derogatis, L. R., & Coons, H. L. (1993). Self-report measures of stress. In L. Goldberger, & S. Breznitz (Eds.), *Handbook of stress* (2nd edition) (pp. 200–233). New York: The Free Press.

Dobson, K. S., & Neufeld, R. W. (1979). Stress-related appraisals: A regression analysis. *Canadian Journal of Behavioral Science, 11,* 274–285.

Dohrenwend, B. S., Dohrenwend, B. P., Dodson, M., & Shrout, P. E. (1984). Symptoms, hassles, social supports, and life events: Problems of confounded measures. *Journal of Abnormal Psychology, 93,* 222–230.

Dohrenwend, B. P., Link, B. G., Kern, R., Shrout, P. E., & Markowitz, J. (1987). Measuring life events: The problem of variability within event categories. In B. Cooper (Ed.), *Psychiatric epidemiology: Progress and prospects* (pp. 103–119). London: Croom Helm.

Erdelyi, M. H. (1974). A new look at the new look: Perceptual defense and vigilance. *Psychological Review, 81,* 1–25.

Fish, T. A. (1986). Semantic differential assessment of benign, threat and challenge appraisals of life events. *Canadian Journal of Behavioral Science, 18,* 1–13.

Folkman, S., & Lazarus, R. S. (1980). An analysis of coping in a middle-aged sample. *Journal of Health and Social Behavior, 21,* 219–239.

Folkman, S., & Lazarus, R. S. (1985). If it changes it must be a process: A study of emotion and coping during three stages of a college examination. *Journal of Personality and Social Psychology, 48,* 150–170.

Folkman, S., & Lazarus, R. S. (1986). Stress processes and depressive symptomatology. *Journal of Abnormal Psychology, 95,* 107–113.

Folkman, S., & Lazarus, R. S. (1988). Coping as a mediator of emotion. *Journal of Personality and Social Psychology, 54,* 466–475.

Folkman, S., Lazarus, R. S., Dunkel-Schetter, C., DeLongis, A., & Gruen, R. J. (1986). Dynamics of a stressful encounter: Cognitive appraisal, coping, and encounter outcomes. *Journal of Personality and Social Psychology, 50,* 992–1003.

Gall, T. L., & Evans, D. R. (1987). The dimensionality of cognitive appraisal and its relationship to physical and psychological well-being. *Journal of Psychology, 12,* 539–546.

Grinker, R. R., & Spiegel, J. P. (1945). *Men Under Stress.* New York: McGraw-Hill.

Horowitz, M. J., Bonanno, G. A., & Holen, A. (1993). Pathological grief: Diagnosis and explanation. *Psychosomatic Medicine, 55,* 260–273.

Horowitz, M. J., Stinson, C., Curtis, D., Ewert, M., Redignton, D., Singer, J., Bucci, W., Mergenthaler, E., Milbrath, C., & Hartley, D. (1993). Topics and signs: Defensive control of emotional expression. *Journal of Consulting and Clinical Psychology, 61,* 421–430.

Horowitz, M. J., Wilner, N., & Alvarez, W. (1979). Impact of event scale: A measure of subjective distress. *Psychosomatic Medicine, 41,* 209–218.

Janis, I. L., & Mann, L. (1977). *Decision making.* New York: The Free Press.

Kanner, A. D., Coyne, J. C., Schaefer, C., & Lazarus, R. S. (1981). Comparison of two modes of stress measurement: Daily hassles and uplifts versus major life events. *Journal of Behavioral Medicine, 4,* 1–39.

Kasl, S. V. (1978). Epidemiological contributions to the study of work stress. In C. L. Cooper & R. Payne (Eds.), *Stress at work* (pp. 3–48). New York: Wiley.

Lazarus, R. S. (1966). *Psychological stress and the coping process.* New York: McGraw-Hill.

Lazarus, R. S. (1991). *Emotion and adaptation.* New York: Oxford University Press.

Lazarus, R. S. (1993). Why we should think of stress as a subset of emotion. In L. Goldberger & S. Breznitz (Eds.), *Handbook of stress* (2nd edition, pp. 21–39). New York: The Free Press.

Lazarus, R. S., Deese, J., & Osler, S. F. (1952). The effects of psychological stress upon performance. *Psychological Bulletin, 49,* 293–317.

Lazarus, R. S., & DeLongis, A. (1985). Stress and adaptational outcomes: The problem of confounded measures. *American Psychologist, 40,* 770–779.

Lazarus, R. S., & Folkman, S. (1984). *Stress, appraisal, and coping.* New York: Springer.

Lazarus, R. S., & Folkman, S. (1987). Transactional theory and research on emotions and coping. L. Laux, & G. Vossel (Spec. Eds.), Personality in biographical stress and coping research. *European Journal of Personality, 1,* 141–169.

Lazarus, R. S., & Launier, R. (1977). Stress-related transactions between person and environment. In L. A. Pervin, & M. Lewis (Eds.), *Perspective in interactional psychology* (pp. 287–327). New York: Plenum.

Lei, H., & Skinner, H. A. (1980). A psychometric study of life events and social readjustment. *Journal of Psychosomatic Research, 24,* 57–65.

Loevinger, J. (1957). Objective tests as instruments of psychological theory. *Psychological Reports, 3*(Monogr. Suppl. 9), 635–694.

Mandler, G. (1975). *Mind and emotion.* New York: Wiley.

McQuaid, J. R., Monroe, S. M., Roberts, J. R., Johnson, S. L., Garamoni, G., Kupfer, D. J., & Frank, E. (1992). Toward the standardization of life stress assessment: Definitional discrepancies and inconsistencies in methods. *Stress Medicine, 8,* 47–56.

Monroe, S. M. (1983). Major and minor life events as predictors of psychological distress: Further issues and findings. *Journal of Behavioral Medicine, 6*(2), 189–205.

Monroe, S. M., & Simons, A. D. (1991). Diathesis–stress theories in the context of life stress research: Implications for the depressive disorders. *Psychological Bulletin, 110,* 406–425.

Monroe, S. M., & Steiner, S. C. (1986). Social support and psychopathology: Interrelations with preexisting disorder, stress, and personality. *Journal of Abnormal Psychology, 95,* 29–39.

Moos, R. H., & Schaefer, J. A. (1993). Coping resources and processes: Current concepts and measures. In L. Goldberger & S. Breznitz (Eds.), *Handbook of stress* (2nd edition, pp. 234–257). New York: The Free Press.

Neale, J. M., Hooley, J. M., Jandorf, L., & Stone, A. A. (1987). Daily life events and mood. In C. R. Snyder & C. Ford (Eds.), *Coping with negative life events: Clinical and social psychological perspectives* (pp. 161–189). New York: Plenum.

Peacock, E. J., & Wong, P.T.P. (1990). The Stress Appraisal Measure (SAM): A multidimensional approach to cognitive appraisal. *Stress Medicine, 6,* 227–236.

Pearlin, L. I., & Schooler, C. (1978). The structure of coping. *Journal of Health and Social Behavior, 19*, 2–21.

Radloff, L. S. (1977). The CES-D Scale: A self-report depression scale for research in the general population. *Applied Psychological Measurement, 1*, 385–401.

Sarason, I. G., Johnson, J. H., & Siegel, J. M. (1978). Assessing the impact of life changes: Development of the Life Experiences Survey. *Journal of Consulting and Clinical Psychology, 46*, 932–946.

Schwartz, J. E., & Stone, A. A. (1993). Coping with daily work problems: Contributions of problem content, appraisals, and person factors. *Work and Stress, 7*, 47–62.

Schwartz, J. E., & Stone, A. A. (in press). Coping with daily work problems: Contributions of problem content, appraisals, and person factors. *Work and Stress.*

Simons, A. D., Angell, K. L., Monroe, S. M., & Thase, M. E. (1993). Cognition and life stress in depression: Cognitive factors and the definition, rating, and generation of negative life events. *Journal of Abnormal Psychology, 102*, 584–591.

Smith, C. A., Ellsworth, P. C. (1985). Patterns of cognitive appraisal in emotion. *Journal of Personality and Social Psychology, 48*, 813–838.

Smith, C. A., Ellsworth, P. C. (1987). Patterns of appraisal and emotion related to taking an exam. *Journal of Personality and Social Psychology, 52*, 475–488.

Solomon, R. C. (1980). Emotions and choice. In A. O. Rorty (Ed.), *Explaining emotions* (pp. 251–281). Berkeley: University of California Press.

Stone, A. A., Greenberg, M. A., Kennedy-Moore, E., & Newman, M. G. (1991). Self-report situation-specific coping questionnaires: What are they measuring? *Journal of Personality and Social Psychology, 61*, 648–658.

Tomaka, J., Blascovitch, J., Kelsey, R. M., & Leitten, C. L. (1993). Subjective, physiological, and behavioral effects of threat and challenge appraisal. *Journal of Personality and Social Psychology, 65*, 248–260.

7

Measurement of Affective Response

Arthur A. Stone

Preliminary Considerations

Affect has a long history in the social and behavioral sciences. It has played a central role in a wide range of theoretical developments and has comprised a large and active field of investigation. One reason for this prominence is that affect and mood are ubiquitous phenomena that are of intrinsic importance to all of us. Even without the benefit of research, each of us knows the relevance of changing affective states and our efforts to achieve the best possible mood. From a clinical perspective, extremes of mood partially define major psychiatric conditions (mania, depression, panic). These observations attest to the significance of affect in our everyday lives and to the importance of understanding the determinants of affect.

A consideration in the writing of this chapter was how to limit what would be covered. One issue centered on the distinctions among other terms related to affect. Mood, affect, and emotion are terms with different meanings depending upon who is defining them (Alpert & Rosen, 1990). Some definitions, including most of the term emotion, refer to a very broad class of phenomena that includes behavioral, cognitive, physiological, and subjective feeling components. A good example is the emotion of fear. Its definition may include escape behaviors, self-statements about the feared object, and a racing heart and sweaty palms. Some assessments have been based on this broad approach to emotion—for example, a questionnaire developed by Wallbott and Scherer (1989). This questionnaire includes sections on a description of the situation experienced, a description of the emotional reaction to the situation, and a section on the control of emotions.

However, from the viewpoint of the stress model presented by Cohen, Kessler, and Gordon (Chapter 1, this volume), the three sections of the Wallbott and Scherer (1984) questionnaire correspond well to the event and appraisals, affects, and coping sections of stress model. In spite of the recent calls for broad definitions of mood (Mayer, Salovey, Gomberg-Kaufman, & Blainey, 1991; Wallbott & Scherer, 1989), such coverage would overlap with many of the other chapters in this book. This chapter is therefore limited to a narrow definition of affect, one concerned with subjective experience. Although mood may be considered a longer lasting state than affect, there is little consensus regarding when affect turns to mood (Alpert & Rosen, 1990). I shall review the assessments that measure both fleeting and longer affect/mood states, and shall use the terms "affect" and "mood" interchangeably throughout the chapter, acknowledging that there is a distinction, however imprecise, between them.

A second consideration was that the history of affect dates to the earliest writings in the field of psychology. This chapter does not attempt to review this history. The interested reader is referred to the many excellent reviews of these topics (e.g., the series of texts edited by Plutchik and Kellerman, see 1989, for instance). This chapter instead takes a very pragmatic approach to the measurement of mood and reviews current issues in its conceptualization and measurement.

A third consideration was whether or not to include trait mood measures. It is self-evident that mood, at least as it is commonly thought of, has the potential to be rapidly fluctuating and responsive to environmental changes. Fluctuations in mood do not, of course, preclude the possibility that there are individuals who have high temporal stability of mood. Some questionnaires have been developed to assess trait mood, often by altering the instructions set used for state measurements so that they are more general, encompassing much longer time periods. There is controversy about this issue (Lorr, 1989) and I shall not review it here. In view of the focus of this chapter on affective *responsivity* and of the methodological issues reviewed below concerning the reporting of mood, I have focused on state assessments.

Rationale for This Measurement Approach

The rationale for measuring affect is its central position in understanding the effects of stress on somatic outcomes and the importance of the construct as an outcome measure in its own right. Cohen, Kessler, and Gordon's model of stress describes a linear transaction and places affective response after appraisal and coping and before physiological and/or behavioral responses. That the five components of the model were placed in sequential order is significant; it implies that each step mediates or influences the process after it. According to the model, unless there is an appraisal of a demand and an inability to cope, there would be no subsequent negative affective response. Likewise, without the affective response, there would be no physiological or behavioral reaction that could lead to disease. It is these relationships that constitute the rationale for studying affective response.

There are several very broad questions about affect and mood that demand our attention. A crucial question for stress theory is, Are affective responses necessary for the effects of events, appraisals, and coping to exert effects on physiological and behavioral processes? Although the current consensus would be, I believe, affirmative, the question has not been adequately addressed in the stress literature. Studies of experimental stressors (mental arithmetic, public speaking), for instance, provide excellent opportunities to explore the mediation of subsequent physiological effects (endocrine and immune, for example) by emotional responses, but few studies have specifically focused on this question (but see Cohen, Tyrrell, & Smith, 1993). A second major question is based on the fact that affect is multifaceted: Are only certain affective responses associated with subsequent physiological and behavioral changes? For example, a standard affective dimension is valence (goodness–

badness) and perhaps only negative mood predicts hormonal response, but positive mood predicts behavioral responses. There is evidence suggesting that even specific moods are related to particular outcomes, reminiscent of Alexander's (1950) specificity theory of emotions (e.g., anger [see Barefoot, 1992] has recently come to light as central to cardiovascular functioning). In addition to intensity and type of affects, questions about the temporal characteristics of affective response have received little attention. Does a short-term affective reaction have the same impact as a larger-term reaction? Are "bursts" of affective activity especially important, because they may be particularly disruptive to physiological function? How do stressor characteristics relate to the temporal patterning of affect?

Another rationale for studying stress-related affect is the considerable literature documenting the relationship between the antecedent components of the model (event occurrence; appraisal and coping) and affective responses. This evidence cuts across the different temporal conceptualizations of stress. We know that measures of stressful events taken at a particular moment are strongly associated with mood at those moments (Shiffman, Gyns, Paty, & Kassel, 1993). There are many studies showing how daily stressful events relate to same-day and later-day moods (see Stone, Neale, & Shiffman, 1993, for a review). Laboratory-based, experimental studies of stressful manipulations dramatically, although not uniformly, influence mood. Not only is there evidence that stressors influence the "normal" range of mood, but also many studies have demonstrated that more severe stressors can affect mood and move it into the "abnormal" range, as in the case of major life events and depression (Brown & Harris, 1978).

In summary, mood is an essential component of the stress model, and there are questions about its relationships with the other domains of variables shown in the model. I have also presented a very brief review of some of the data demonstrating that stress is strongly related to mood reactivity and that these relationships exist at several temporal levels.

Kinds of Questions That the Measure Can Answer

Some of the other constructs considered in this book have very specific functions in the study of stress. Major life events are a useful way of conceptualizing stress for studying outcomes with relatively long periods of actions, for example, in the development of cancer or hypertension. Daily events are useful for investigating more rapidly changing processes, for example, those that could result in respiratory illnesses. And extremely brief experimental stressors have their place in the study of rapidly changing physiological processes. Unlike these specific stress measures that match up well with specific outcomes, affect is a construct that a researcher could consider studying—perhaps *should* study—in virtually all types of stress research. As we have seen, it is a major component of the model and, depending upon the researcher's conceptualization of affect, can serve a study in several ways.

One way that affect can be conceptualized is as an *outcome* measure. A researcher might be interested in how the frequency of daily events impacts on mood. In this case, affect can be viewed as a general indicator of the state of the organism, with positive moods suggesting a state of good function and negative moods a perturbed state (Morris, 1989). It is also notable that most "well being" measures include an affective component; such measures are typically used as outcomes. Alternatively, mood can serve as a *clinical outcome,* as one component of a depressive or manic diagnosis (these syndromes are composed of nonaffective signs and symptoms as well). A different conceptualization of affect is as a measure of *stressor severity.* An example is the researcher who studies a single major life event, say that of job loss, and employs affective response as a measure of the severity of the event. This investigator might hypothesize that subjects with greater affective responses, remembering that job loss event is constant for all subjects, are the ones who will have serious consequences of the loss. The assumption made in this scenario is that mood is a final pathway of sorts: all of the effects of the stressful event, subjects' personality and supports, and other potential moderators of the event's impact are integrated and have a final psychological effect that emerges in affect. As mentioned above, it has not been demonstrated that this assumption is correct, but this is one type of question that mood has been used to address in stress research.

Choosing an Appropriate Measure

General Considerations in the Measurement of Mood

Structure of Mood

A controversial issue in this research area concerns the structure of mood. Two basic positions have been advanced concerning its structure: the *specific affects approach* and the *dimensional approach.* The first approach is characterized by the belief that there are many different types of mood, each with different, although related, characteristics and response patterns (Izard, 1977; Tomkins, 1963). Affect "labels" resulting from this view are thought to be separate and independent of one another. Measures that result in scales with labels such as happiness, sadness, fear, remorse, and so on, have been derived from the specific affects approach. An important hypothesis pertaining to this approach is that specific patterns of psychophysiological response are thought to be associated with different mood states. The second approach is based on the notion that there are a few, usually two, "core" dimensions of mood; specific moods are thought to be combinations of the "basic" dimensions. One very influential dimensional conceptualization is the *circumplex model* of mood, where two dimensions are crossed, forming perpendicular diameters of a circle. Particular moods are thought to lie around the circumference of the circle. This view of affect is analogous to color hues: Colors can be conceptualized as a color circle, where discrete color labels (red, blue) are viewed as mixtures of more primary colors.

Historically, the specific affects approach to mood was the first to be developed. Factor analyses of mood adjective checklist responses have yielded many solutions, and the resulting factors defined the discrete mood labels that are the hallmark of this approach. On the other hand, dimensional approaches have used the same or similar mood descriptors, but start the search for factors with several assumptions about the structure of mood that differ from the specific affect approach. I shall spend more time discussing the dimensional position, because it currently appears to be the more dominant of the two approaches, although facial expression research, which is part of the specific affect tradition, is also very influential (and is briefly described below).

Larsen and Diener (1992) have summarized the assumptions underlying the circumplex approach: that some emotions are more similar to each other than are other emotions; that all emotions can be formed by combinations of two basic dimensions; that affect can be described in a circular space; and that the order of affects along the circle is determined by the model. There are two camps regarding the naming of the basic circumplex dimensions: the Negative Affectivity (NA) and Positive Affectivity (PA) camps (Watson & Tellegen, 1985) and the Unpleasant/Pleasant and High/Low Activation camps (Russell, 1980). Figure 7.1 describes the circumplex with the axes and affect labels inserted around the circle's circumference. The issues involved in the choice of one set of labels versus another are complex, but an important point is Russell's argument that Watson's NA and PA dimensions are confounded with activation.

A compelling position advanced by Larsen and Diener (1992) is that it is impossible to choose one set of basic dimensions over another either on psychometric grounds or by reference to mood's association with external criteria. These authors discuss the circumstances under which the circumplex model may be appropriate and in which it may not: They conclude that "[d]imensional approaches seem to gloss over many interesting features of emotion" p. 8). For example, the fact that anger and fear are located close to each other on the circumplex suggests great similarity, yet in contradistinction to this implication, the two emotions are quite different. In other words, dimensional approaches miss much of the richness of affective life and, furthermore, do not convey differences about how different affects are represented in nonsubjective realms (e.g., facial expressions, psychophysiological arousal). This position suggests that there is room for both specific affect and dimensional approaches in stress research.

Frequency Versus Intensity of Mood

An important yet often overlooked issue in the measurement of mood concerns how the frequency with which people experience moods relates to the intensity with which moods are experienced (Diener, Larsen, Levine, & Emmons, 1985). In a series of longitudinal studies, subjects used different instruction sets for completing a mood adjective checklist. The instructions varied in terms of the period of time covered (ranging from momentary to the entire day) and included an instruction for mood to be reported after especially strong emotional experiences. Results showed that reports of mood intensity were not correlated with reports of the frequency of moods, suggesting that affect intensity and affect frequency are two discrete components of mood. If frequency is measured over relatively long periods of time, then PA and NA will correlate negatively. If PA and NA intensities are measured, then

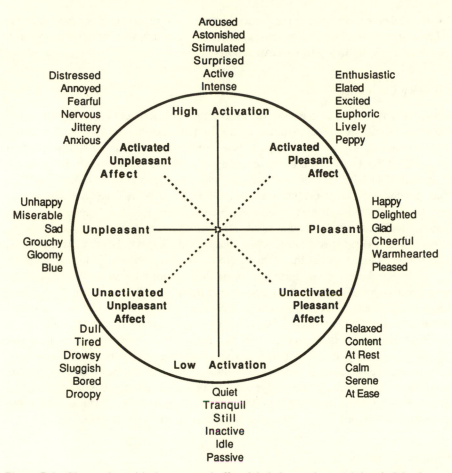

Figure 7.1. Circumplex with the axes and affect labels inserted around the circle's circumference.

the scales will be independent. However, if one moves to very small periods of time, the relationship between the intensity of PA and NA reverses: A strong inverse relationship is observed, and the relationship is strongest for the smallest reporting periods (e.g., momentary). Related to the concepts of intensity and frequency is that of the duration of moods, and, unfortunately, very little is known about the duration of particular mood "episodes."

These concepts and relationships appear important for understanding the structure of mood, yet to date they have had very little effect on the development of mood instruments. What is clear is that the structure of mood is different when subjects report their moods over long periods of time as compared to when they report their moods over short periods. I think that a strong argument can be made for not having subjects summarize their mood over long periods of time, on two grounds: First, it is not known exactly how subjects complete the task of summarizing mood over time, and they probably do not all do it in the same way. Second, it is known that memory bias is more prone to influence retrospective reports (see the next section). Although somewhat limited by their specificity, momentary mood reports appear to be more accurate than retrospective reports.

Memory Capability

An extremely important question for assessing mood, especially when one is considering differing temporal instruction sets, is, How capable are people of reporting their moods, and over what time periods are such reports valid? We know that there are memory issues involved in the reporting of stressful events, and it is reasonable to assume that brief, subjective states are even more prone to memory distortions than are discrete life events (see DeVries, 1992). There is also evidence demonstrating that ability to recall past mood (and events) is influenced by mood state at the time of the recall task, making mood-congruent memories more accessible (e.g., Bower, 1981; Donner & Young, 1991; Teasdale & Fogarty, 1979). These findings suggest that shorter, rather than longer, mood reporting periods should be assessed.

Although there is no definitive work on the how long a period is acceptable for retrospective recall of mood, two studies bear directly on this question. Thomas and Diener (1990) examined this issue by comparing reports of momentary mood with retrospectively recalled mood. They concluded that "if any general statement can be made, it is that retrospective reports of one's emotional experiences over time tend not to be extremely accurate" (p. 295). In a study that was restricted to mood recall over a single day, Hedges, Jandorf, and Stone (1985) found that momentary mood reports correlated with end-of-day reports of mood in a pattern showing that end-of-day reports captured significant information about daily mood. These results suggest that reports of 24-hour mood are valid at least to some degree.

Specific Response Biases

Reports of mood are prone to the same biases as are other psychological constructs, so it is necessary to consider how typical biases influence mood reports. Three biases have been explored. The first bias is a common problem for behavioral research: social desirability. This bias stems from a tendency of some subjects to complete assessments so that they appear in a positive light. One measure of this tendency, the Marlow–Crowne Social Desirability Scale, has been employed by some mood researchers. The second potential bias that has received attention is a tendency for subjects to use high numbers in their responses to mood questionnaires. This kind of bias could create artifactual between-subject associations among truly unrelated measures. Diener et al. (1985) addressed this potential problem in a study of affect intensity by creating an index of the tendency to use high ratings, which was used as a covariate in later analyses. In the Diener et al. (1985) study, control of these bias did not affect the main findings. The third bias, resulting from the structure of response scales, has been discussed by Meddis (1972). Using the Nowlis mood scale as an example, Meddis argued convincingly that the "cannot decide" response option, which is placed between "slightly" and "definitely not," does not fall on the same "intensity" dimension as the remaining response options.

This has the potential to create response artifacts. Other work (e.g., see Svensson, 1978) has extended and confirmed this work. There is a fourth potential bias that may be relevant in applications where mood is assessed repeatedly. The specialty of psychological testing has been sensitive to the possibility that responses to a test administered on one occasion influence responses on subsequent occasions and have thus devised parallel test forms to reduce such bias. This has not been the case in the development of mood assessments. The issue appears especially salient when mood is self-assessed many times in relatively short periods; many study design have subjects make multiple mood ratings, sometimes with only a few minutes separating the administrations. Such designs may foster stereotypic responding, leading to higher associations among measurements and lower variance in mood over time.

Influences on Mood the Investigator May Wish to Control

In addition to biases in the measurement of mood, there are other well-documented influences that an investigator employing a mood measure should consider. The potential impact of these factors on mood will vary by study design and purpose. Day of the week has very reliable influences on mood (Larsen & Kasimatis, 1990; Stone et al., 1985) and may need to be controlled for either by statistical procedures (covariance analyses) or by experimental means (balancing groups, e.g., by day of the week). A related influence (and one that may account for the day-of-week effect) is the association between daily happenings and mood (see Stone et al., 1993); events have overall daily effects and, from a momentary approach, even quicker, more fleeting effects on mood (Shiffman et al., 1993; Stone et al., 1992). Likewise, there may be other diurnal patterns of mood, possibly related to biological cycles. Caffeine, nicotine, and alcohol also have transient effects on mood (Rogers, Edwards, Green, & Jas, 1992). There is also some evidence that menstrual cycles affect mood (Warner & Bancroft, 1988). A researcher contemplating the inclusion of mood in their study should consider which of the above factors could influence their mood reports, and take action to control these effects. For example, controlling the intake of caffeine and nicotine prior to a laboratory study or monitoring menstrual cycle in naturalistic studies are two ways an investigator could address these issues.

Choosing an Appropriate Measure

Surveying the Measures

A large number of mood assessment instruments are available to the stress researcher; there is, however, much similarity among them. For this reason, I present mainly descriptions of the most widely accepted mood measures, and a few especially interesting measures, rather than attempt an exhaustive survey of the measures. I also describe more fully the most recent approaches to mood assessment. (The reader interested in a very detailed presentation of additional instruments is referred to MacKay, 1980). It is important to note that I have depicted the scales and instructions to participants as presented in the articles describing the scales. It is possible, and in many cases perfectly reasonable, for another investigator to alter these instructions to suit the needs of a study or to eliminate a potential bias that might occur, for instance, from the use of an ambiguous or too lengthy reporting period. The disadvantage of doing this is, of course, that the validity and reliability information presented for that scale is best thought of as specific to the instructions used for obtaining the information. There are several general methods that have been used to assess affect and mood. I present the methods and specific instruments in terms of increasing complexity of procedures. (Table 7.1 presents a summary of the assessments reviewed.)

Single-Item Measures

The simplest method and the one with perhaps the highest degree of face validity is to ask subjects to rate their mood on a multipoint scale. The question might target a dimension of mood, such as negative mood or anger, and the response key could range from 0/not at all, to 7/extremely. Another response scale that is often used with single-adjective descriptors of moods or very broad mood descriptions (e.g., "positive" or "negative" mood) is called a visual analogue scale (VAS). Instead of using a number to indicate the degree to which the mood was experienced, a subject checks a point along a horizontal line in an effort to avoid some of the responses associated with the use of numerical scales (e.g., tendency not to check the extremes). The ends of the scale are often marked with adjectives such as "extremely" or "not at all." One study that compared a VAS scale to an adjective checklist scale concluded that VAS methods might be most appropriate for measuring global concepts of affect, whereas other questionnaire methods were required for more specific information (Morrison & Peck, 1990). In experimental studies, single-item assessments of "tension" have been used. In daily studies, ratings of mood throughout the day have been assessed by means of a single item with "good" and "bad" as endpoints. Virtually any single mood construct (unipolar or bipolar) can be assessed in this manner.

An alternative mood assessment relies on graphical presentation of cartoon-type faces with various emotions depicted on them. Usually several faces are presented and a subject is instructed to pick the one that best describes their mood. Lang (1980) has developed an assessment using two sets of faces: One set depicts the dimension of pleasure–displeasure, and the other set depicts arousal. This instrument has been used in the context of laboratory studies. Subjects indicate which faces represent their moods.

A very different graphical assessment is to have a subject place a checkmark in a grid representing different intensities and qualities of mood. Two measures employ this method.

Affect Grid This is a single-item assessment of the circumplex model of mood using the dimensions favored by Russell (1980). A subject is presented with a graphic that is composed of a 9 × 9 matrix. Affect descriptors are placed at each corner of the square and at the midpoint of each side. Starting at the upper-left-hand corner and

Table 7.1. Measures of Affect

Scale	Number of Items	Item Style Response Categories	No. of Factors or Dimensions
Affect Balance Scale	10	phrases describing feeling states (5 positive, 5 negative) rated yes or no	2
Profile of Mood States	65	Adjectives rated on 4- or 5-point scales	6
Multiple Affect Adjective Checklist	132	adjectives that are checked or not checked	3
Clyde Mood Scale	44	adjective rated on 4-point scales	6
Nowlis Mood Adjective Checklist	36	adjectives rated on 4-point scales	12
Differential Emotions Scale–IV	36	descriptions of feelings rated on 5-point scales	10–12
Semantic Differential Mood Scale	35	pairs or bipolar adjectives rated on 5-point scales	5
Positive and Negative Affect Schedule	20	adjectives rated on 5-point scales	2
Thomas & Diener	9 or 12	4 or 6 positive and 5 or 6 negative affect terms rated on 7-point scales	2
Activation–Deactivation Adjective Checklist	8	adjectives rated on 4-point scales	4
Forced-Choice Adjective Checklist	33	selection of one of a pair of adjectives that best describes mood	variable
Neutral Word Ratings	6	neutral words rated on 7-point scales for pleasantness	1
Visual Analogue Scale	1	respondent marks mood on a 100-mm line with descriptive terms at ends	1
Circular Mood Scale	1	circle outlines with adjectives and marked for current mood	2
Affect Grid	1	adjectives around a 9 × 9 matrix which is marked for cell best describing mood	2

moving clockwise, the descriptors are as follows: stress, high arousal, excitement, pleasant feelings, relaxation, sleepiness, depression, and unpleasant feelings. A subject then places a check into the cell of the grid that best describes his or her current mood (or other instructions referring to specific stimuli). Data presented by the developers of the instrument (Russell, Weiss, & Mendelsohn, (1989) suggests that it compares favorably with other, more cumbersome assessments of pleasure and arousal, and several other studies have shown that the scale responds as expected to environmental manipulations. Importantly, the authors report that subjects were willing to complete the instrument many times.

Circular Mood Scale. This scale is similar to the Affect Grid. Subjects mark how they are feeling along the perimeter of a circle that has mood adjectives every 45 degrees (at eight places; Jacob et al., 1989). The adjective placement is very similar to the circumplex model of mood; for example, unhappy and happy are opposite each other. A difference between this and other dimension approaches is that the dependent variable is formed from the trigonometric relationships among the dimensions represented by the adjectives. In addition, subjects also check which of discrete moods (following the specific affect approach) they are experiencing. The moods are anger, fear, sadness, joy, and "other."

Adjective Checklists

By far the most common method of assessing mood is with mood adjective checklists. In these assessments, a number of adjectives describing emotional states are listed, and subjects are instructed to indicate if the moods described by the adjectives reflect their feelings. There are many variations on this basic theme, involving different response scales associated with adjectives, different sets of adjectives, and different instruction sets for the adjectives.

Response scales that have been used range from those asking for a "yes" or "no" answer as to whether or not the adjective applies to current mood, to 9-point ratings. The sets of adjectives used in mood adjective checklists vary markedly both in content and in the number of adjectives making up a scale. Some checklists are as short as three adjectives, whereas other, more comprehensive scales contain over 100 adjectives. The adjectives comprising checklists are quite varied, depending upon the method of scale development used. Some checklists of the same or similar lengths may have very few adjectives in common. Finally, the instruction sets for responding to adjectives differ markedly, most notably in terms of period of time that the subject is supposed to be characterizing in their report. Some checklists ask for an appraisal of immediate affect, whereas other checklists request how subjects "generally" or "usually" feel; the latter is an attempt to measure a traitlike quality of mood, as opposed to more rapidly fluctuating momentary mood. Between these temporal extremes are many variations, including *daily* mood instruction sets, those asking about the last *hour,* and those asking about the last *5 minutes.*

Nowlis Mood Adjective Checklist (MACL). This was one of the first instruments employing adjectives to assess mood. The MACL was developed in 1965 by Nowlis and Green on the basis of earlier (1957) work utilizing factor analyses of 130 adjectives. The more popular short form of the MACL contains the following 36 adjectives: angry, clutched-up, carefree, elated, concentrating, drowsy, affectionate, regretful, dubious, boastful, active, defiant, fearful, playful, overjoyed, engaged in thought, sluggish, kindly, sad, skeptical, egotistic, energetic, rebellious, jittery, witty, pleased, intent, tired, warmhearted, sorry, suspicious, self-centered, and vigorous. Each adjective is rated according to how the subject feels at the time the word is read, by the subject's circling of one of four response options: (1) definitely felt it, (2) slightly, (3) cannot decide, (4) definitely not. Twelve factors have been found for the MACL: Aggression, Anxiety, Surgency, Elation, Concentration, Fatigue, Social Affection, Sadness, Skepticism, Egotism, Vigor, and Nonchalance. An even shorter form of the scale, based on factor analyses of the 36 items has been developed. This 12-item checklist uses the original response key and has been shown to yield a positive and negative engagement scale (Stone, 1981).

Multiple Affect Adjective Checklist (MAACL). Originally developed by Zuckerman and Lubin (1965), this is also a widely used mood checklist. It was designed to assess clinical aspects of depression, anxiety, and hostility with 132 adjectives, although only 89 are required to measure the three dimensions. Hundreds of studies have used the MAACL with a wide range of subjects in a variety of research designs (see Gotlib & Meyer, 1986). An issue for users of this measure is the high intercorrelation among the three scales, leading some to create a single scale score (Howarth & Schokman-Gates, 1981). Recently, however, Gotlib and Meyer (1986) factor-analyzed the MAACL and reported not one, but two factors. These factors were labeled positive affect and negative affect. Interestingly, in this study subjects rated current mood, unlike prior studies that had subjects rate mood over the previous week.

Profile of Mood States (POMS). Originally developed for evaluating psychiatric patients, this instrument was first called the Psychiatric Outpatient Mood Scale (McNair, Lorr, & Droppleman, 1971). However, the name that ultimately stayed with the instrument, and that has the same acronym (i.e., POMS), was introduced the following year. It is composed of 65 adjectives that are rated for how the subject felt during the past week. This instruction set has raised the possibility that the POMS is more of a trait measure than a state measure (Howarth & Schokman-Gates, 1981). Many investigators have, however, shortened the reporting period to suit their needs. Six factors have been identified: Tension–Anxiety, Depression–Dejection, Anger–Hostility, Vigor–Activity, Fatigue–Inertia, and Confusion–Bewilderment.

Activation–Deactivation Adjective Checklist. Presented by Thayer in 1967, this scale was intended to assess momentary activation and arousal. The short form of the scale contains the following adjectives: energetic, lively, active, vigorous, full-of-pep, sleepy, drowsy, tired, wide-awake, tense, clutched-up, fearful, jittery, intense, still, at-rest, calm, and quiet. Each adjective is rated with one of four options: (1) definitely feel, (2) feel slightly, (3) cannot decide, and (4) definitely do not feel. Consistent with prior studies, a recent factor analyses of the adjectives (Thayer, 1986) resulted in several factors: General Activation factor (called Energy), Deactivation–Sleep (Tiredness), High Activation (Tension), and General Deactivation (Calmness).

Clyde Mood Scale. This checklist was developed to measure mood changes in psychiatric patients (Clyde, 1963). The response scale for the adjectives has four options: (1) not at all, (2) a little, (3) quite a bit, and (4) extremely. The scale contains 44 adjectives; six dimensions, based on factor analyses, have been identified: Friendly, Aggressive, Clear-Thinking, Sleepy, Unhappy, and Dizzy.

Differential Emotions Scale. There are several versions of this questionnaire, but all propose to measure 10 basic emotions identified from the facial expression research (joy, surprise, sadness, anger, disgust, contempt, fear, shame/shyness, and guilt [Izard, Dougherty, Bloxom, & Kotsch, 1974]). Each concept is rated on a multi-point scale for how the subject currently feels.

Positive Affect–Negative Affect Schedule (PANAS). The PANAS scales were developed by Watson, Clark, and Tellegen (1988) to tap two major dimensions of mood: positive affectivity (PA) and negative affectivity (NA). One of the distinguishing aspects of the scale's development was its emphasis on maintaining an independence between PA and NA by selecting "pure" markers for each scale. Ten adjectives make up each scale. For PA the adjectives are attentive, interested, alert, excited, enthusiastic, inspired, proud, determined, strong, and active. For NA they are distressed, upset, hostile, irritable, scared, afraid, ashamed, guilty, nervous, and jittery. Each adjective is rated on a 5-point scale labeled (1) very slightly or not at all, (2) a little, (3) moderately, (4) quite a bit, and (5) very much. Seven different sets of temporal instructions have been examined for these adjectives, including (1) right now, (2) today, (3) past few days, (4) past week, (5) past few weeks, (6) past year, and (7) on average (Watson et al., 1988). Internal consistencies for both scales with all of the instruction sets have been above .84, and the correlation between NA and PA has been very low, regardless of which instruction set was used.

Semantic Differential Mood Scale. An alternative scaling of mood is available based on the semantic differential approach (Osgood, Suci, & Tannenbaum, 1957). Lorr and Wunderlich (1988) developed a scale in an attempt to reduce response bias, to create bipolar scales, and to assess positive moods. Subjects are instructed to check which of a pair of affect descriptors (e.g., cheerful–dejected) best describes their feeling state at the moment. The scale was developed with high school students, and five factors were identified: Elated–Depressed, Relaxed–Anxious, Confident–Unsure, Energetic–Fatigued, and Good-Natured–Grouchy.

Russell's Adjective Checklist. As mentioned above, Russell (1979, 1980) developed a mood assessment based on a two dimensional bipolar conception of affect. The adjectives making up this scale are shown in Figure 7.1.

Brief, Momentary Mood Checklists. Thomas and Diener (1990) recorded momentary emotions with four positive words (happy, joyful, pleased, enjoyment/fun) and five negative words (depressed/blue, unhappy, frustrated, angry/hostile, worried/anxious). These were rated on 7-point response scales (0 = not at all; 6 = extremely much). Another scale for use on a momentary basis was developed by Shiffman and colleagues (1993). Two features of this scale are notable: First is the *response scale.* Designed for administration with a handheld computer, it employs the following response key: "YES!!" "Yes." "No?" and "NO!!" Each item is assigned a score ranging from 1 to 4. Second, in addition to several adjectives being rated, scores from two-dimensional ratings of *Overall feeling* (response key: very bad, bad, neutral, good, very good) and *Energy level* (response key: very low, low, neutral, high, very high) are also included in the Valence and Arousal scales derived from the instrument. An even briefer momentary mood assessment has been used by Marco and Suls (1993). They assessed negative emotions with a three-item scale composed of the adjectives tense, angry, and unhappy. Each adjective was rated on a 6-point response scale, anchored on the extremes with (1) not at all and (6) very much.

Forced-Choice Adjective Checklist. This is an adjective checklist, but one with a twist. Instead of rating the degree to which an adjective applies to one's mood, pairs of adjectives are presented to subjects (e.g., boastful–egotistical) with the instruction that they are to choose the adjective that most applies to their mood during the previous 2 hours (Wendt & Cameron, 1961). Developed for use in drug studies, this response format may avoid some of the response bias found in the usual method of responding to adjective checklists.

Global Mood Scale. This is a scale developed for use in coronary heart disease patients to assess emotional distress and fatigue (Denollet, 1993). I present this scale for two reasons. First, it was developed especially for cardiac patients because the author felt that a specialized scale was required to capture mood adequately in such patients; that other mood scales might "miss" mood that was relevant for these patients; and that the mood scale needed to be sensitive to rehabilitation interventions. Second, the scale was developed in the Netherlands using adjectives based on the PA/NA two-dimensional model of Watson and Tellegen (1985), demonstrating the cross-cultural impact of that formulation of mood. A 5-point response key ("not at all" to "extremely") was used with the instruction to respond to each mood "experienced lately." In the group of cardiac patients, the PA and NA scales correlated as predicted by the authors with shortened POMS, state anxiety, and well-being scales, yet did not correlate with social desirability.

Observational Methods

These methods rely on someone else's, other than the subject's, description of a subject's moods.

Observer-Rater Mood. Another approach to the measurement of mood and affect is to have someone other than the person being studied to evaluate or judge mood. A straightforward method of employing observers in the assessment of mood is to have individuals who themselves have contact with the subject rate the subject's mood with a modified self-report scale (Stone, 1981). Although this procedure does remove the assessment from the arena of the self-report, it has other problems. The most pressing one is that it is not necessarily clear on what basis is the observer judging the subject's mood: for example, what aspects of the subject's behavior are they using in the ratings? Are they listening to the subject describe his or her mood and then simply using that description for the ratings? If so, one could question whether the procedure is really nothing more than a distorted version of the self-report method. Or, is the observer making use of other information concerning the expression of the subject's mood? At this point it is premature to recommend this mood assessment for general usage; considerable research needs to be conducted to arrive at a better understanding of the meaning of these assessments.

Facial Expressions. Another observational method is based on classifying subjects' facial expressions. Although this type of measurement does not fall into the subjective mood domain described earlier (instead falling into a broader definition of emotion), it is a very influential approach and so is briefly mentioned. Unlike the first observation method, facial expression procedures have received considerable attention and are well worked out. This is not the place to go into the conceptual reasons for studying affect via facial expressions (for a review, see Eckman, 1992). It must be noted, though, that the facial expression approach to emotion has had a large impact on basic issues in this field including (1) validity of the dimensional versus specific mood theories and (2) elucidation of the physiological pathways involved in emotion and mood. From a purely measurement perspective, the advantages of the approach are that it avoids the problems inherent in self-report technologies and that reliable procedures have been developed for implementing these measurements. There are several drawbacks, most of them logistical, concerning the use of facial expressions assessments in many types of studies. Nevertheless, there are times when a stress researcher working in the laboratory may wish to employ this method.

Other Techniques

Affect Balance Checklist. Developed by Bradburn (1969) in the late 1960s, this was one of the earliest mood measures. Subjects are presented with 10 statements pertaining to their mood, to which subjects respond with a "Yes" or "No" answer. It can be scored for positive and negative affects and for affective balance. A recent study examined these scales in an elderly sample (Kempen, 1992).

Neutral Word Ratings. A newly developed approach to the measurement of affect is based on research demonstrating the pervasive effects of mood states on judgment processes (Forest, Clark, Mills, & Isen, 1979). The idea is that when subjects are asked to judge the pleasantness/unpleasantness of neutral words (e.g., south, rock, coffee), it is not the word per se, but their own affective states that influence their ratings. Kuykendall, Keating, and Wagaman (1988) have shown that mood induction procedures affect neutral word ratings in expected directions and that mood adjective checklist assessments of naturally occurring mood states correlate with neutral word ratings. Although this is a relatively unexplored procedure, it is likely that such procedures avoid some of the response issues that are of concern for other, more obvious mood instruments. The techniques developed by Kuykendall et al.

(1988) present subjects with 10 words that, on the basis of semantic differential ratings, have been shown to be neutral. Each word is rated on a 7-point scale, anchored at one end with pleasantness and at the other end with unpleasantness. The words used are south, rock, solitude, coffee, ice, censorship, bear, weight, beard, and down.

Projective Techniques. Another method of measuring mood and one that has not received much attention in the stress literature, is based on projective techniques. For example, emotion-oriented scales have been based on Rorschach test responses. These techniques take the "broad" view of emotion mentioned above and therefore are not discussed here. The interested reader is referred to Kellerman (1989).

Psychometric Properties

For many psychological measures, *test–retest reliability* is one of the hallmarks of a psychometrically sound instrument. Test–retest reliability assesses the degree to which a measurement is stable over time, and moderate-to-high test–retest reliabilities are expected for stable characteristics of persons such as intelligence and personality (varying, of course, on the interval between testings). This is not a reasonable criterion for mood measures, because they are conceptualized as states that have the ability to fluctuate rapidly (Russell, Weiss, & Mendelsohn, 1989). Very substantial test–retest reliabilities would contradict that proposition, and would actually suggest that a measure was tapping more traitlike properties of individuals (although it is certainly possible that for certain people or conditions, high test–retest reliability could be observed). Therefore, the relatively low between-occasion reliabilities that have been observed (MacKay, 1980) are just what one would expect. It is possible to compute another reliability index based on the tendency for responses to a scale item to co-vary) with scales that have more than one question, to produce an assessment of the scale's *internal consistency.* It is not surprising that moderately high levels of internal consistency are consistently observed, when one considers the extensive factor-analytic procedures that have been used to develop many mood scales.

All of the mood scales, with the exception of the Global Mood Scale, the Semantic Differential Mood Scale, and the Experience Sampling Method (ESM) scales, mentioned above, have sufficient validity data to be considered for use in a stress study. Each measure has been shown to change in expected directions when respondents experience environmental changes that were expected to be associated with affect (see MacKay, 1980, for a summary of some of this information). Actually, the data are quite overwhelming in terms of the sheer number of studies that have utilized mood measures and that have shown predicted changes in the scales. One should, I think, be especially careful in reviewing the validity information available for single-item scales, since internal reliability data are not available for them.

Population Appropriateness

With the exception of graphical forms of mood scales, the assessments that have been discussed depend heavily on subjects' interpretations of affect-related descriptors and are in self-report format. This has several implications for the use of these measures in different populations. First, subjects must be literate and able to understand the distinctions among the various mood descriptors. The questionnaires may not, therefore, yield dependable results when used with individuals who have, for any of several reasons, poor semantic comprehension. For the same reason, the usefulness of the questionnaires may be compromised in samples whose primary language is not the same as the one used in the development of the instrument. Although these subjects might be able to read the words, their comprehension may not be sufficient for producing valid mood reports. Second, many of the instruments mentioned above have been translated into other languages. Although it is not unusual for psychological instruments to be translated, this is an especially precarious process for mood scales. Words referring to a particular affect in one language may not translate into a word describing the same affect in another language. (It may be that certain affects are not even recognized in particular cultures.) Therefore, procedures to ensure appropriate comparability (e.g., extensive back translation) are important in this research area. Third, young children may not have the understanding of affect descriptors required to complete mood questionnaires; however, these scales have been used with high school students. The graphical approaches may also be used with children. Fourth, for some of the reasons mentioned above, the elderly may have difficulty with some mood measures. These issues are discussed by Schultz, O'Brien, and Tompkins (1994), who also recommends scales for use with this population.

Logistical Issues

Length of Administration

Mood measures are very flexible in this regard, especially if the researcher is willing to adopt a dimensional approach. If only a very brief amount of time is allocated to mood measurement, then a researcher may wish to consider a VAS. Such scales are very quick, have considerable face validity, and have been used in a variety of studies, often laboratory-based ones. A subject can easily complete a single VAS scale in a matter of a few seconds. However, it is probably worth taking a few more seconds, if possible, to move to a small adjective checklist or one of the graphical measures. These types of scales are also face valid, but have the advantage of being able to assess more than a single dimension of mood (although that is also possible with VAS methods) and, in regard to the checklists, give the user the opportunity to test a scale's internal consistency and create a more reliable measure. For the researcher who can allocate a longer time to mood assessment, many of the instruments mentioned above could be used. Even checklists with 40 items require less than 10 minutes for an average subject to complete.

Special Equipment or Requirements for Subjects

No special equipment is required for subjects to complete mood assessments. However, there may be designs and/or subject samples where the paper-and-pencil format typically used presents a problem. Subjects who are unable to read (because they are blind or illiterate) or write will not be able to complete the measures. Likewise, the researcher conducting a telephone interview will not be able to use these measures in the standard way. While I know of no research directly comparing written versus verbal administration of adjective checklists, there seems to be no obvious reason for not adapting the scales for verbal administration. One issue in the administration of mood scales either by telephone or to subjects who are unable to read the questionnaire is that it is difficult for subjects to use the unfamiliar rating scales (e.g., remembering the range of the scale, or what the endpoints are labeled with, etc.). One solution to this problem is to employ response options that are familiar to most subjects. An example of this technique is to have subjects rate of mood adjectives on a 0- to 100-point "percentage scale," a metric that most of us encounter in everyday life (Shiffman, personal communication). Other considerations may also arise in particular populations. For example, the elderly may have limited concentration spans, and thus it is appropriate that shorter scales be employed with this group. Likewise, this group may tend to present themselves in a particularly positive light and/or they may have limited experience completing rating scales, suggesting that special care be paid to the instructions developed for this group (see Schultz et al., 1993).

Evaluation of Mood Measures

It is a difficult task to answer the question, Which of the reviewed mood measures is the best? I concur with the conclusion of Howarth and Schokman-Gates's (1981) review of the existing mood report measures: "It becomes difficult to provide specific instrument recommendations as to preferring this rather than that scale" (p. 437). A stress researcher will have to choose a mood measure based on a number of factors, and those factors will limit the domain of measures. Before discussing some general considerations about the selection of a mood scale, I shall comment on the development of specific mood scales.

Development of Idiosyncratic Mood Scales

It is clear that there is an abundance of affect and mood scales available to the stress researcher. Many of these have a well-proven track record, having been used in literally hundreds of studies (e.g., the Nowlis and MAACL). One may reasonably ask, then, Under what conditions should a new mood scale be developed? I presented the Global Mood Scale as an example of a scale that was developed for a specific purpose; however, I did not comment on the rationale or the validation of the questionnaire. I do not believe that the rationale was particularity sound: What was the evidence that the items on extant mood scales were not "relevant" to cardiac patients? And should we think of a mood scale as being poor if it does not show the effects of a treatment? This seems to me to be a step backwards, defeating the work to establish valid mood scales and diminishing the possibility of comparing results across studies. On the validation of the scale, it is odd that a direct comparison between the new scale and extant scales was not carried out, especially in light of the strong associations between new and older scales. Although I have no particular issue with the resulting GMS instrument, I think that new scales developed without a strong rationale should be accepted with much caution.

Selecting a Scale

While it may not be possible to say which of the reviewed is "best," I think that it is possible to draw some general conclusions regarding the selection of a mood scale. These recommendations are certainly not hard and fast, but may provide a starting point for thinking about scale selection.

1. Some investigators may wish to obtain general information about subjects' moods. For example, do major life events have an effect on subsequent mood? Although this is a perfectly reasonable question, the cross-sectional study designs often employed in life event studies may lead researchers to measure mood using an adjective checklist with instructions for subjects to respond to how they "usually" feel. The studies reviewed above suggest that this methodology may be problematical, mainly because responses to that instruction set may be more indicative of personality than of mood and because memory and summary biases are possible with retrospective mood reports. An alternative procedure is to measure much shorter periods, for example, a momentary period, the preceding few hours, or even an entire day. A single measurement will, of course, reflect more proximal influences on mood (e.g., current events) as well as the hypothesized effects of past life events. This will increase "error" variance and may reduce statistical power to levels unacceptable for the study design. To achieve more reliable estimates of average mood, and thus reduce bias from personality factors and recall processes, investigators can assess mood on several occasions and average the results (Epstein, 1979).

2. Related to the previous point is how mood fits into the hypotheses of a study. If there are no specific hypotheses about how moods will be related to antecedent or subsequent processes—for example, in studies where mood is not a primary variable—then I suggest that an investigator be conservative and choose a mood assessment that yields a smaller number of general factors than a large number of specific moods. The rationale for this recommendation is that, given the lack of hypothesized relationships, the investigator has a greater risk of observing spurious associations with many mood scales than with a smaller number. If, for example, an association is found between the predictor measure and one or two of eight mood scales, the investigator is left in the difficult position of explaining why those associations were found. Of course, there is an argument for the opposite position: If a study is exploratory, then why not explore many moods? The bottom line for both of these positions (although more for the second one) is, of course, that

whatever associations are found in a single study, conclusions should not be accepted until they have been replicated.

3. Although most of the mood measures presented have an abundance of validity information, some scales have considerably less information. It is advisable to use scales that have demonstrated the capacity to change in situations similar to those the investigator will be studying. For instance, the laboratory researcher should look to those scales that have been shown to be responsive to rapid alterations in conditions. A researcher interested in an end-of-day mood summary should probably look to other scales.

4. If the study design allows, it is probably better to select a larger adjective checklist than a shorter one. Certainly, studies assessing mood in naturalistic protocols many times per day cannot afford to impose 75 mood questions at each assessment. On the other hand, many studies can ask subjects to answer 10 to 12 mood adjectives, given how quickly these instruments can be completed. It is a well-known statistical reality that internal-consistency reliability increases as the number of items comprising a scale increases; however, the relationship is not linear, and the return in increased reliability decreases once a scale reaches a certain size. Furthermore, others have commented that longer scales are not necessarily better (Burisch, 1984).

5. An investigator who chooses the dimensional approach to mood measurement is immediately faced with the issue of which dimensions to measure. Although the question has by no means been resolved, I am concerned with the interpretation of the unipolar positive and negative affect factors, and find the bipolar intensity and arousal factors more interpretable. However, it is true that very few studies have compared these positions, and therefore, my concern is founded on conceptual, rather than on firm empirical, grounds.

Future Directions

Much progress has been made in the measurement of mood since the 1960s, and thus the researcher interested in its measurement is hardly at a loss from, and is perhaps even overwhelmed by, the number of assessments. Despite this reality, it is very difficult to evaluate the strengths and weaknesses among these assessments, and one must ask why that is. Part of the answer is that mood is an intrinsically difficult construct to measure. Nevertheless, it is striking that so few comparisons among the various questionnaires have been conducted. Investigations of multiple mood scales could clarify the distinctions among different set of adjectives, different response options, and different ways of scoring the tests. Such investigations do not appear to be particularity complex; yet they seem necessary for advancing the field. My suspicion is similar to those of other reviewers of mood measurement: that different ways of measuring mood will be differentially associated with outcome and response biases. The field of mood measurement needs to know which methods should be used for which questions. Such studies would also, I believe, advance our basic understanding of mood. A second issue concerns the meaning of response scales used in mood assessments. Recent developments in scale construction, including, for example, Item Response Theory (Hambleton & Swaminathan, 1985), could be used to test whether not different groups of subjects (defined by age, ethnicity, country of origin, etc.) use response scales in similar ways. The third issue that I believe deserves the attention of mood researchers is that of understanding the meaning of retrospective mood assessments. Although some progress has been made in this area, the basic question is still unanswered. In what reporting period do memory biases and processes related to summarizing mood come into play? Are there ways in which questions can be asked to allow subjects to report past mood more accurately? Studies of momentary mood assessments and retrospective assessments, along with creative debriefing tasks and/or probes about the retrospective mood reports, could help to answer these questions. For example, designing scales that assess frequency, duration, and intensity of moods over some period (say, 24 hours) could be validated against point-assessed mood reports.

Acknowledgments

The author thanks Dr. Saul Shiffman for his suggestions on an early draft of this chapter. I also thank Dr. Richard Schultz for his helpful suggestions and for allowing me to modify his table of mood measures. Finally, I thank the editors of this volume for their comments.

References

Alexander, F. (1950) *Psychosomatic medicine*. New York: W. Norton.

Almagor, M., & Ben-Porath, Y. (1191). Mood changes during the menstrual cycle and their relation to the use of oral contraceptive. *Journal of Psychosomatic Research, 35,* 721–728.

Alpert, M., & Rosen, A. (1990). A semantic analysis of the various ways the terms "affect," "emotion," and "mood" are used. *Journal of Communication Disorders, 23,* 237–246.

Barefoot, J. (1992). Developments in the measurement of hostility. In H. Friedman (Ed.), *Hostility, coping, and health* (pp. 13–31). Washington, DC: American Psychological Association.

Bower, G. (1981). Mood and memory. *American Psychologist, 36,* 129–148.

Bradburn, N. (1969). The structure of psychological well-being. Chicago, IL: Aldine.

Brown, G. W., & Harris, T. (1978). *Social origins of depression: A study of psychiatric disorder in women*. New York: Wiley.

Burisch, M. (1984). You don't always get what you pay for: Measuring depression with short and simple versus long and sophisticated scales. *Journal of Research in Personality, 18,* 81–98.

Clyde, D. (1963). *Manual for the Clyde Mood Scale*. Coral Gables, FL: University of Miami.

Cohen, S., Tyrell, D. A., & Smith, A. P. (1993). State and trait negative affectivity as predictors of objective and subjective symptom response to viral infections. University of Pittsburgh.

Denollet, J. (1993). Emotional distress and fatigue in coronary heart disease: The Global Mood Scale (GMS). *Psychological Medicine, 23,* 111–121.

DeVries, M. (1992). *The experience of psychopathology*. Cambridge, UK: Cambridge University Press.

Diener, E., Larsen, R., Levine, S., & Emmons, R. (1985). Intensity and frequency: Dimensions underlying positive and negative affect. *Journal of Personality and Social Psychology, 48,* 1253–1265.

Donner, E., & Young, M. (1991). *Mood variability and accuracy of recall for depression.* Paper presented at the annual meeting of the Society for Research in Psychopathology.

Eckman, P. (1992). Facial expressions of emotion: New findings, new questions. *Psychological Science, 3,* 34–38.

Epstein, S. (1979). The stability of behavior: I. On predicting most of the people much of the time. *Journal of Personality and Social Psychology, 37*(7), 1097–1126.

Forest, D., Clark, M., Mills, J., & Isen, A. (1979). Helping as a function of feeling state and nature of the helping behavior. *Motivation and Emotion, 3,* 161–169.

Gotlib, I., & Meyer, J. (1986). Factor analysis of the Multiple Affect Adjective Check List: A separation of positive and negative affect. *Journal of Personality and Social Psychology, 50*(6), 1161–1165.

Hambleton, R. K. & Swaminathan, H. (1985). *Item response theory.* Boston: Kluwer-Nijhoff.

Hedges, S., Jandorf, L., & Stone, A. (1985). Meaning of daily mood assessments. *Journal of Personality and Social Psychology, 48,* 428–434.

Howarth, E., & Schokman-Gates, K. (1981). Self-report multiple mood instruments. *British Journal of Psychology, 72,* 421–441.

Izard, C. (1977). *Human emotions.* New York: Plenum.

Izard, C., Dougherty, F., Bloxom, B., & Kotsch, W. (1974). The differential emotions scale: A method of measuring subjective experience of discrete emotions. Unpublished manuscript (Department of Psychology, Vanderbilt University).

Jacob, R., Simons, A., Manuck, S., Rohay, J., Waldstein, S., & Gatsonis, C. (1989). The circular mood scale: A new technique of measuring ambulatory mood. *Journal of Psychopathology and Behavioral Assessment, 11,* 153–173.

Kellerman, H. (1989). Projective measures of emotion. In R. Plutchik & H. Kellerman (Eds.), *Emotion: Theory, research, and experience, Vol. 4: The measurement of emotions* (pp. 187–203). San Diego, CA: Academic Press.

Kempen, G. I. J. M. (1992). Psychometric properties of Bradburn's Affect Balance Scale among elderly persons. *Psychological Reports, 70,* 638.

Kuykendall, D., Keating, J., & Wagaman, J. (1988). Assessing affective states: A new methodology for some old problems. *Cognitive Therapy and Research, 12,* 279–294.

Lang, P. J. (1980). Behavioral mtreatment and biobehavioral treatment: Computer applications. In J. B. Sidowski, J. H. Johnson, & T. A. Williams (Eds.), *Technology in mental health care delivery.* Norwood, NJ: Albex.

Larsen, R., & Diener, E. (1992). Promises and problems with the circumplex model of emotion. In M. S. Clark (Ed.) *Review of Personality and Social Psychology, Vol 14: Emotional and social behavior* (pp. 25–59). Newbury Park: Sage.

Larsen, R., & Kasimatis, M. (1990). Individual differences in entrainment of mood to the weekly calendar. *Journal of Personality and Social Psychology, 58*(1), 164–171.

Lorr, M. (1989). Model and methods for measurement of mood. In R. Plutchik & H. Kellerman (Eds.), *Emotion: Theory, research, and experience, Vol. 4: The measurement of emotions.* New York: Academic Press.

Lorr, M., & Wunderlich, R. (1988). A semantic differential mood scale. *Journal of Clinical Psychology, 44,* 33–36.

MacKay, C. (1980). The measurement of mood and psychophysiological activity using self-report techniques. In I. Martin & P. Venables (Eds.), *Techniques in psychophysiology* (pp. 501–562). New York: Wiley.

Marco, C., & Suls, J. (1993). Daily stress and the trajectory of mood: Spillover, response assimilation, contrast and chronic negative affectivity. *Journal of Personality and Social Psychology, 64,* 1053–1063.

Mayer, J., Salovey, P., Gomberg-Kaufman, S., & Blainey, K. (1991). A broader conception of mood experience. *Journal of Personality and Social Psychology, 60,* 100–111.

McNair, D., Lorr, M., & Droppleman, L. (1971). *Psychiatric Outpatient Mood Scale.* Boston, MA: Psychopharmacology Laboratory, Boston University Medical Center.

Meddis, R. (1972). Bipolar factors in mood adjective check lists. *British Journal of Social and Clinical Psychology, 11,* 178–184.

Morris, W. (1989). *Mood: The frame of mind.* New York: Springer.

Morrison, D., & Peck, D. (1990). Do self-report measures of affect agree? A longitudinal study. *British Journal of Clinical Psychology, 29,* 395–400.

Nowlis, V., & Green, R. (1957). *The experimental analysis of mood.* Technical Report, Office of Naval Research: Contract No. Nonr-668(12)

Osgood, C., Suci, C., & Tannenbaum, P. (1957). The measurement of meaning. Urbana, IL: University of Illinois Press.

Plutchik, R., & Kellerman, H. (1989). *Emotion: Theory, research, and experience, Vol. 4: The measurement of emotions.* New York: Academic Press.

Rogers, P. J., Edwards, S., Green M. W., & Jas, P. (1992). Nutrition influences on mood and cognitive performance: The menstrual cycle, caffeine, and dieting. *Proceedings of the Nutritional Society, 51,* 343–351.

Russell, J. (1979). Affective space is bipolar. *Journal of Personality and Social Psychology, 37,* 345–356.

Russell, J. (1980). A circumplex model of affect. *Journal of Personality and Social Psychology, 39,* 1161–1178.

Russell, J., Weiss, A., & Mendelsohn, G. (1989). Affect grid: A single-item scale of pleasure and arousal. *Journal of Personality and Social Psychology, 57,* 493–502.

Shiffman, S., Gyns, M., Paty, J., & Kassel, J. (1993). Minor daily events and momentary mood: A validation study.

Stone, A. (1981). The association between perception of daily experiences and self- and spouse-related mood. *Journal of Research in Personality, 15,* 510–522.

Stone, A. A., Hedges, S. M., Neale, J. M., & Satin, M. S. (1985). Prospective and cross-sectional mood reports offer no evidence of a "blue Monday" phenomenon. *Journal of Personality and Social Psychology, 49,* 129–134.

Stone, A., Neale, J., & Shiffman, S. (1993). How mood relates to stress and coping: A daily perspective. *Annals of Behavioral Medicine, 15,* 8–16.

Stone, A., & Shiffman, S. (1992). Reflections on the intensive measurement of stress, coping, and mood, with an emphasis on daily measures. *Psychology and Health, 7,* 115–129.

Svensson, E. (1978). Mood: Its structure and measurement. *Goteborg Psychological Reports, 8,* 1–19.

Teasdale, J., & Fogarty, S. (1979). Differential effects of induced mood on retrieval of pleasant and unpleasant events from episodic memory. *Journal of Abnormal Psychology, 88,* 248–257.

Thayer, R. (1986). Activation–deactivation Adjective Check List: Current overview and structural analysis. *Psychological Reports, 58,* 607–614.

Thomas, D., & Diener, E. (1990). Memory accuracy in the recall of emotions. *Journal of Personality and Social Psychology, 59,* 291–297.

Tomkins, S. (1963). *Affect, imagery, and consciousness, Vol. 11: The negative affects.* New York: Springer.

Wallbott, H., & Scherer, K. (1989). Assessing emotion by questionnaire. In R. Plutchik & H. Kellerman (Eds.), *Emotion: Theory, research, and experience, Vol. 4: The measurement of emotions* (pp. 55–82). New York: Academic Press.

Warner, P., & Bancroft, J. (1988). Mood, sexuality, oral contraceptives, and the menstrual cycle. *Journal of Psychosomatic Research, 32,* 417–427.

Watson, D., Clark, L., & Tellegen, A. (1988). Development and validation of brief measure of positive and negative affect: The PANAS Scales. *Journal of Personality and Social Psychology, 54*(6), 1063–1070.

Watson, D., & Tellegen, A. (1985). Towards a consensual structure of mood. *Psychological Bulletin, 98,* 219–235.

Wendt, G., & Cameron, J. (1961). Chemical studies of behavior: V. Procedures in drug experimentation with college students. *Journal of Psychology, 51,* 173–211.

Zuckerman, M., & Lubin, B. (1965). *The Multiple Affect Adjective Check List.* San Diego, CA: Educational and Industrial Testing Service.

PART IV

The Biological Perspective

The biological tradition focuses on the activation of biological systems that are particularly responsive to physical and psychological demands. Prolonged or repeated activation of these systems is thought to place persons at risk for the development of a range of both physical and psychiatric disorders. Two interrelated systems that are viewed as the primary indicators of stress response are the autonomic nervous system, especially the sympathetic–adrenal medullary system (SAM), and the hypothalamic–pituitary–adrenocortical axis (HPA). The measures of SAM activation addressed in this volume include the cardiovascular and endocrine systems; measurement of HPA activation is primarily endocrine. Relations between stress and other biological systems, especially the immune system, have recently received considerable attention as well.

Chapter 8, dealing with the measurement of stress hormones, provides an overview of what is known about the hormones that are most commonly investigated in studies of stress and provides information about measurement of several hormones and neuropeptides that appear most often in the stress literature. Chapter 9, reviewing cardiovascular measurement, summarizes the response of different components of the cardiovascular system to stressors, discusses cardiovascular reactivity, and provides recommendations on the measurement of various components of the system. Finally, Chapter 10, discussing immune measurement, provides an overview on the role of the immune system and of the components of immune response. It describes several measures of immunity commonly used in stress research and many of the practical complexities involved in immune measurement in studies of stress.

There are, of course, other biological indicators of the stress response that are not discussed in this volume, for two reasons: space constraints and our interest in choosing measures that we felt would be of interest to the broadest possible audience. For example, both skin conductance and muscle tension are often used as biological markers of stress in laboratory (and recently field) studies. Moreover, along the pathway between the central nervous system and effects on disease outcome are a myriad of biological intermediates that may be explored. In studying diseases of the gastrointestinal system, for example, hormonal measures could include those measured in Chapter 8, but also gastrointestinal peptides and measures of gastric motility. As technology develops in areas such as brain scanning, the variety and specificity of biological measurement of the stress responses in the central nervous system will increase. We look forward to the contributions that these factors will make to research on the stress response and its relationship to physical and mental health.

8

Measurement of Stress Hormones

Andrew Baum and Neil Grunberg

The broad applicability or pervasiveness of stress has made it a useful construct in a variety of areas of study, ranging from human performance and mood to eating, drug use, and the etiology and progression of disease. As a result, a wide range of stress outcomes have been studied, and changes related to this syndrome have been considered in models of diverse phenomena. One consequence of this breadth of study is diffusion of the hazy meaning of the construct: Does stress consist of sympathetic nervous system-driven physiological and biochemical changes, or is it simply emotional turmoil? Are measures of blood pressure or urinary catecholamines equivalent to measures of negative mood and performance deficits? If, in fact, stress is a broad and poorly defined process, how can neuroendocrine markers be reasonably incorporated into its study? What does it mean when neuroendocrine, physiological, and self-report measures of stress are not highly intercorrelated? These and other issues have punctuated the expansion of stress study to include measures of neuroendocrine activity and discussion of exactly what these measures are.

The first use of hormones as explanatory mechanisms in models of stress was Cannon's (1914) depiction of epinephrine as a primary mobilizer of the emergency response, also known as fight-or-flight. Cannon reasoned that many of the peripheral physiological changes that occurred during stress, including changes in skeletal and visceral muscle tone, were achieved by the action of some substance released during stress by the glands sitting atop the kidneys (adrenals). He was able to demonstrate the differential actions of this substance, called sympathin then (epinephrine now), by applying blood drawn from stressed animals to muscle preparations in the laboratory. He observed varied responses to stimulation by blood from stressed animals that were characterized as adaptive changes designed to assist the organism in dealing with sources of stress.

Selye's (1976) model of stress also was based on neuroendocrine changes, but he focused on a different system. He initially discovered that stressors such as heat, x-rays, and injection of extracts all caused the same three physiological changes: (1) hypertrophy of the adrenal glands, (2) involution of the thymus and lymphoid tissue, and (3) ulceration in the GI tract. These changes were considered to be consequences of extended and unusually intense activity in the pituitary–adrenal cortical system and the resulting massive release of corticosteroids. Selye also

viewed these hormonally stimulated changes as adaptive for the organism's surviv-
al. Since these early investigations, models of stress that focus on neuroendocrine
mediators have become more common. The number of hormones (chemical substances with specific regulatory functions) and peptides (low-molecular-weight amino acid compounds) thought to be important during stress has increased dramatically, due in part to technological developments that permit detection and
differentiation of more substances. At the same time, awareness of stress outcomes,
particularly mental and physical health outcomes, has expanded, and investigators
have begun to study links between stress and a great array of pathology, necessitat-
ing inclusion of measures of more biochemicals and activity in more neuroendocrine systems.

This chapter was written to provide an overview of the most common hormonal
measures of stress: the information they provide, the conditions under which they
may be used, and the costs and benefits of such use. Rationale for using endocrine
markers, either as outcomes or mediating conditions, are also discussed, and the
general place of endocrine activity in models of stress is considered. Use of such
measures broadens one's perspectives on stress and other biobehavioral processes
and may provide critical information in investigations of stress effects on health,
mood, and behavior.

Stress and Neuroendocrine Activity

Though definitions and depictions of stress vary, all or most posit an integrated
biological response pattern during or after stressor exposure. The activity of the
endocrine system in the periphery and of neurotransmitter systems throughout the
body represents a basic, underlying layer upon which other aspects of this response
are built. Some changes appear to be alerting or alarm-oriented and serve to facilitate the initiation of the stress response. Other neuroendocrine changes are supportive of a general systemic response, by increasing availability of energy, potentiating
the response, and/or by facilitating mobilization or recovery. In addition, many of
these neuroendocrine changes are thought to be mechanisms by which stress may
affect other systems or contribute to pathogenesis and disease.

The nature of bodily changes during stress may be cast in a coping framework if
one views stress as a process that unfolds as organisms encounter, appraise, and
respond to situations that pose threat, challenge, loss, or demand (see Chapter 1).
That is, when an event or situation is stressful, a cascade of hormonal changes
occurs that appears to work either to motivate or to support coping with the stressor.
Upon recognition of threat or demand, sympathetic nervous system (SNS) activity
increases with a corresponding increase in catecholamine release and whole-body
activation. This initial response, identical to Cannon's emergency response, can be
experientially aversive (e.g., rapid heart rate, increased sweating, GI distress) and
can motivate the organism to act either to remove or to reduce this tension, minimize its effects, or eliminate its source (e.g., Baum, 1990). At the same time, SNS
arousal, together with other neuroendocrine changes, increases the availability of

glucose, redistributes blood to skeletal muscle and away from the skin and other vulnerable areas, and otherwise supports whatever coping responses are applied. The activating functions of SNS activity alone serve to heighten the strength and speed of these responses, and other changes are intimately involved in metabolic regulation during stress.

The list of neuroendocrine changes thought to be associated with stress has grown very long. We do not consider all of them, but focus on those that have been used more commonly over the past few decades and those that currently appear to be particularly useful in explaining how stress affects disease processes or how stress-related changes interact with other systems. The catecholamines, epinephrine (E) and norepinephrine (NE), as well as corticosteroids (CO) and their metabolites, have been the most commonly studied neuroendocrines in stress research and are of considerable importance in the initiation and regulation of stress responses. At the same time, these adrenal hormones appear to be the primary mechanisms whereby stress affects other systems and pathogenesis. Endogenous opioid peptides (EOPs) have received a considerable amount of attention as well; they have been implicated in pain response, stress-related drug use, eating behaviors, affective responses, and immune system changes. The CNS neurotransmitter serotonin (5-hydroxytrypta-mine, 5-HT), its precursors (e.g., tryptophan), and its primary metabolites (e.g., 5-hydroxyindoleacetic acid, or 5-HIAA) have also become the focus of many recent studies addressing a variety of psychological phenomena related to stress, including depression and other affective disorders, obsessive-compulsive disorders, eating disorders, and drug abuse. Finally, a number of factors and hormones, including growth hormone, nerve thyroid-stimulating hormone, insulin, and others have been measured as they change in blood concentration during stress. Of these, thyroid hormones have been studied as correlates of mental health problems, prolactin has been studied as an immunomodulator, and growth hormone has been studied in several contexts.

Catecholamines

One of the more common and frequently used measures of hormonal activity in stress is the assessment of circulating or excreted levels of epinephrine and nor-epinephrine. The catecholamines, also including dopamine and a few other variants, are synthesized from the amino acid tyrosine in brain and adrenal medullary chro-maffin cells, sympathetic nerves, and sympathetic ganglia. They have a wide range of metabolic effects and stimulatory effects on most organ systems when released into circulation (dopamine is a precursor of norepinephrine, some of which is synthesized into epinephrine). The adrenal medullae are ordinarily activated during SNS arousal, and the release of epinephrine and norepinephrine from the adrenals and sympathetic neurons elsewhere in the body heighten and extend SNS arousal already achieved through neural stimulation. Synthesis and release of cate-cholamines are regulated by feedback systems, neuronal depolarization, Ca^{2+}-dependent processes, prostaglandins, vasoactive amines, polypeptides, and acetylcholine. Furthermore, levels of catecholamines in circulation, particularly of NE, may reflect the amount released, NE not reabsorbed by synaptic receptors, or a combination of these and other factors.

Plasma and Urinary Catecholamines

Circulating levels of catecholamines can be measured by assaying blood samples, but the half-life of epinephrine and norepinephrine is very short (1–3 minutes; Berne & Levy, 1983), and turnover or decay of circulating levels can occur within a minute or two. As a result, measures from blood reflect acute states. Determinations of levels of E and NE in the blood reflect recent levels of SNS activity and should not necessarily be considered as indicating anything more than a transient or momentary state. For studies of SNS reactivity to acute laboratory stressors, such momentary snapshot measures are useful because they reflect rapid changes and can depict catecholamine levels at repeated points in time. For study of the effects of chronic stress, these blood measures are not particularly useful because of considerable fluctuation and intraindividual variability in E and NE levels.

Levels of catecholamines also can be estimated by measuring free E and NE or metabolites of E and NE in the urine. A small but consistent fraction of circulating catecholamines are excreted as free (i.e., unmetabolized) E and NE, and although these are not necessarily accurate measures of absolute levels, they are useful in comparisons among individuals or longitudinally in the same subjects. Examination of free urinary E and NE allows differentiation of the two catecholamines as does analysis of some metabolites such as metanephrine and normetanephrine. The alternative, measuring other metabolites of E and NE in the urine, does not permit differentiation because the two are metabolized into the same compounds, primarily vanillylmandelic acid (VMA) and methoxyhydroxyphenylglycol (MOPG). Assessments of these metabolic products provide good estimates of overall catecholamine activity but do not permit evaluation of separate levels of E and NE.

Measurement of urinary E and NE or assay of metabolites provide longer, more stable indices of SNS activity than do blood measures. Because urinary catecholamine and metabolite levels are excreted slowly over a period of hours during which the bladder fills, these measures constitute average estimates of SNS arousal over the period between voids (the margin of error being about an hour or less). For studies of ambient or chronic stress, 24-hour urine samples have traditionally been used. These long-term urine samples provide an estimate of ambient or average levels through the day, combining resting and active or upset experiences. If one has reason to suspect episodes of elevated hormones during the day, one can ask subjects to keep each urine void separate and labeled as to time and activity. If all urine is collected, total daily and/or hourly excretion rates can be determined across the entire day. However, compliance with 24-hour urine sampling can be a problem, and 15-hour samples collected overnight (e.g., 6 PM to 9 AM) may suffice. This approach does not require subjects to collect urine while at work or on the go during the day, when compliance is ordinarily more difficult and activity confounds more likely.

Because these urinary measures are more sluggish and insensitive to immediate cues, they have not been particularly useful in studying acute stress responses. Even taken at the end of an experimental session, E and NE levels in a urine sample would likely reflect several hours of experience and, unless collected 45 minutes or more after the stressor, might not reflect any stressor experience at all. A double-void technique has been used in order to make some use of urine-based catecholamine measures in acute stress studies. Subjects arrive at the laboratory and void before entering the study so as to remove all urine (and catecholamines or metabolites) reflecting prior experiences. A second sample, taken at the end of the session, is thought to reflect the subjects' experience after this void and could be construed as an index of SNS arousal over the entire session.

Two problems with this approach are immediately apparent. First, there will inevitably still be some effect of pre-void experience present in the system after the first void. In addition, the averaging effect of urine samples will diffuse SNS increases over a 2-hour period, diluting and obscuring effects of the stressor. Because it takes some unspecified and variable amount of time for catecholamines to be degraded and urine to form and reach the bladder, it is unclear where the cutoff would be for what experience is "included" in a particular urine sample. If something stressful occurs as a subject is on his or her way to a session, say 30 minutes before, effects of this experience may not be "excreted" in the first void but may affect the second sample. In addition, the results of urinary assessments of E and NE after a 2-hour laboratory session frequently show little change; for example, 15 minutes of stressor experience may be lost in 105 minutes of sedentary activities or rest that make up the remainder of a laboratory session.

Assays

There are several ways to express catecholamine data. They are typically reported either as a concentration per unit of urine or blood or as a total volume of catecholamines excreted over 24 hours or on an hourly basis. The use of concentrations predominates in studies using blood sampling and analysis of catecholamines in circulation. Although estimates of total E and NE can be made from such determinations, it is not possible to measure the exact total blood volume at a particular time and it is difficult to draw enough blood to be able to estimate hourly rates of secretion. With urine samples, the choices are broader; concentrations have been reported, and in double-void laboratory studies there are often insufficient data to allow adjustment by urine volume. However, for long-term samples (e.g., 15, 24 hours) measurement of volume of urine collected is useful to provide some reassurance that the participant complied with urine collection instructions and to allow correction for total urine production. Measuring total urine volume allows one to modify concentrations of E and NE and permits estimates of total amounts of catecholamines produced. If a participant in a study based on 24-hour sampling shows a concentration of NE of 20 ng/ml and excreted 1000 ml during the 24-hour period, total daily production can be estimated at 20×1000 or 20,000 ng. In this case, the hourly rate would be 20,000 ng/24 hr = 833 ng. If this amount of urine and this concentration of NE were obtained with 15-hour samples, procedures for adjusting the data by volume would be more complex. Because the entire 24-hour day was not captured and because of the possibility that rates of catecholamine release vary over the 24-hour day, total day estimates cannot be made from 15-hour samples. Instead, hourly rates would be calculated by multiplying the concentration by the volume and dividing by 15 (in this case, $20 \times 1000/15 = 1333$ ng). But this hourly value is not necessarily comparable to a 24-hour collection analysis.

Currently, there are two principal techniques used to measure levels of catecholamines in biological fluids; radioenzymatic assay (REA) and high-performance liquid chromatography (HPLC). REA of biological samples for catecholamines exposes each sample to the enzyme catechol-*O*-methyltransferase (COMT) in the presence of a compound (tritiated *S*-adenosylmethionine or SAM) that donates a radioactive label to the primary metabolite of each catecholamine molecule present (see Durrett & Zeigler, 1980). Subsequent steps in the REA procedure clean up the samples to concentrate the radiolabeled catecholamines. In addition, techniques are applied that separate the primary metabolites of E, NE, and DA from one another so that the amount of each of these catecholamines can be determined. In contrast, HPLC exposes biological fluids to extremely high pressure to separate and measure amounts of catecholamines present in comparison with known standards. This technique relies on physical properties (e.g., molecular weight and size) or chemical characteristics (e.g., ionic change, types of chemical bonds) of the chemicals being analyzed. Concentrations of the separated catecholamines usually are determined by electron capture. Both techniques have been validated extensively and are reliable. The primary differences are that the REA technique requires the use of radioisotopes, several trained technicians, and a scintillation counter. HPLC does not require radioisotopes but does require at least one highly skilled technician and an expensive and sensitive HPLC system.

Blood and urine samples collected for assessment of catecholamines are subject to oxidative and degradation processes that can result in loss of detectable catecholamines and inaccurate results. These natural processes must be prevented and/or arrested if proper interpretation of these samples is to be made. The simplest way to do this is to freeze samples as quickly as possible, at least at −20°C for urine samples and −70°C for plasma. In addition, particularly when refrigeration and freezing are not immediately feasible, a preservative can be added which will stop these degradation processes. For example, in collection of 15 or 24-hour urine samples, one cannot always guarantee that subjects will keep their collections in a refrigerator or on ice. Antioxidants such as sodium metabisulfite and hydrochloric acid have been used, and recent experience suggests that for many assays, these preservatives are comparable (Cohen, personal communication). Our experience has been that sodium metabisulfite is an excellent preservative for urine sampling; it is not caustic, holds samples with minimal oxidative loss for 24 hours without refrigeration, and is easy to use. Hydrochloric acid can cause injury if spilled and therefore may not be as useful for urine sampling. For blood samples, blood should be drawn directly into heparized or EDTA-treated tubes, spun, and frozen without preservative. It is essential that the particular assay being used be identified prior to initiation of procedures.

Considerations for Use of Catecholamine Measures

The cost of catecholamine assays varies widely depending on whether one uses a commercial laboratory. Currently, costs run between $30 and $120 per sample, with costs in one's own lab with a technician and equipment at the low end. Use of urine-based samples requires compliance with collection regimens, and if subjects do not provide full samples, problems can result. Total volumes and creatinine clearance can be used as indicators of the adequacy of extended urine collection. In addition, these are two major issues central to any decision to use measures of catecholamines in stress studies. First, one must determine if one needs to use them. As a simple manipulation check, that is, to determine whether a stressor was associated with heightened SNS arousal, these measures are probably unnecessary. They may show a slightly stronger effect than do physiological measures such as blood pressure (BP), but they are expensive, require blood draw or long-term urine collection, and add several layers of difficulty to even routine stress investigations. For cases in which the confirmation of effects of a stressor is the question, simpler measures of BP or HR usually suffice. However, when one is interested in the mechanisms by which stress affects immune function or interacts with other bodily systems, or if one is studying the convergence and concordance of stress-related changes, measures of E and NE are important. Without them, putative mechanisms remain unstudied and unclear. With them, study of the means by which stress affects us may be better understood.

This argument also holds for studies in which the points of stress-induced disease effects are of interest. If one is interested in heart disease or diabetes, the hormonal changes accompanying stress may help to explain etiology or disease course and may be important disease mechanisms. Study of them appears to be essential for drawing conclusions about stress and health. And, while this logic holds for measures of other endocrine changes as well, the conditions under which study of catecholamine changes is necessary or useful should be considered. For example, catecholamines, and NE in particular, are very sensitive to movement or activity. Simply standing up from a seated position causes dramatic increases in circulating catecholamines. This means that activity before and during blood draws or urine sampling must be controlled, and exercise and physical activity must be measured in chronic stress studies. In addition, catecholamines are affected by drug use, consumption of caffeinated beverages and alcohol, and eating many common foods as well as cigarette smoking. These consumatory behaviors must be controlled for or measured when catecholamines are used as dependent variables.

Corticosteroids

Another common hormonal measure of stress is the assessment of circulating or excreted corticosteroids (CO). Produced and secreted by the adrenal cortex, CO are secreted in increased quantities during and after exposure to some stressors. This release is part of the systemic arousal of the hypothalamic–pituitary–adrenalcortical axis (HPA) initiated by release of corticotropic releasing hormone (CRH) by the hypothalamus. CRH stimulates the pituitary to produce adrenocorticotropic hormone (ACTH), which in turn, elicits corticosteroid release from the adrenals. During stress, this production is enhanced, and larger quantities of primarily glucocorticoids are released in pulsatile fashion (in bursts). The adrenal cortex may be thought of as being on or off: Between pulses of CO release the cortex is quiescent. This refractory period allows the glands to replenish CO just released.

Glucocorticoids are synthesized in the zona fasciculata of the adrenal cortex. Cortisol is the primary glucocorticoid for humans, whereas corticosterone, produced in small amounts in humans, is the primary glucocorticoid in some animal species. As with the catecholamines, a small amount of cortisol is excreted as free cortisol in urine and can be measured by radioimmunoassay (see below). Most of the excretory product of cortisol is in the form of metabolites; cortisol and corticosterone are metabolized primarily in the liver and excreted as tetrahydrocortisol and tetrahydrocortisone. These 17,21-dihydroxy-20-ketone (17-OHCS) compounds were the basis for early measurements of the urinary 17-hydroxycorticoids (which constitute up to half of total cortisol secretion).

Cortisol and corticosterone release show clear diurnal rhythms. While dependent on ACTH, cortisol release appears to peak at about 8 AM, with a low point around midnight. However, cortisol is released in pulsatile bursts, and measurement over a 24-hour day can reveal 15 or more pulse releases of cortisol, with the largest during the early morning. This pattern may be a result of a learned association with eating, as the essential functions of cortisol include maintenance of glucose production from protein as well as facilitation of fat metabolism. Other corticosteroid functions include anti-inflammatory actions and immune, renal, and muscle function regulation. To a large extent the actions of cortisol and other glucocorticoids are catabolic in that they are focused on pulling energy out of stored reserves and facilitating arousal and action. They also appear to interact with other hormones (e.g., glucagon) and with arousal of the SNS through synthesis of E.

Plasma and Urinary Cortisol

The HPA axis is not as rapidly responsive as is the SNS. Arousal of this system may take several minutes and increases in circulating levels of cortisol are sluggish and not immediately manifest. The half-life of cortisol (about 70 minutes) is longer than that of catecholamines. This means that different timing issues become important. If one is interested in measuring circulating catecholamines and cortisol as a function of an acute stressor, baselines for catecholamines cannot be taken right after venipuncture because the SNS-stimulating effects of the venipuncture will affect E and NE right away. As a result, one should wait until the effects of the needle stick have passed, perhaps 15–20 minutes afterward, before drawing blood for baseline catecholamine levels. However, effects of venipuncture on cortisol might not be evident for at least 10–15 minutes after venipuncture, so an immediate postdraw sample would provide a useful baseline for cortisol. Using the 15 to 20 minute post–needle stick measure for baseline cortisol might result in sampling at the point at which the effects of venipuncture are *most* evident. Using the same blood draw to provide baseline samples for both catecholamines and cortisol would be problematic. Of course, using an immediate post-stick blood sample as baseline assumes that presession activity and distress are controlled (e.g., by a rest period).

For many animal studies in which hormonal indices of stress are used, corticosterone is often more useful than are catecholamines because it is less transient and responsive to immediate stimuli. In most of these studies, animals are sacrificed and blood is collected for assay. E and NE are usually affected by the sacrifice procedures. Effects of sacrifice or collection procedures on corticosterone are not apparent in collected samples because they do not change as rapidly.

Historically, the HPA axis and measures of corticosteroid release have been described as mechanisms of stress and used as indices of stress far longer than have measures of catecholamines, in part because assays of CO and it's metabolites were developed earlier than were measures of other hormones. In addition, Selye's (1976) model of stress focused on corticosteroid release, and Mason (e.g., 1975a) in his rebuttal of Selye and his extension of stress research also focused initially on adrenal cortical compounds. However, it was largely due to Mason's work that multiple measures of different endocrine systems are now measured. Mason (1975b) argued that stress was manifested in a number of systems and that simultaneous measurement of corticosteroids, catecholamines, insulin, thyroid hormones, and other endocrines was the best way to investigate stress.

Salivary Cortisol

For the most part, the literature described above was based on plasma and urine measures of cortisol. A relatively new development in the study of cortisol and integration of HPA axis activity in stress research and clinical research has been the development of reliable measurements of cortisol in saliva. The advantages of such a technological advance are clear: Saliva samples can be obtained without stressful and invasive procedures such as venipuncture and do not require a phlebotomist or other medical personnel for data collection. This technique also offers a more acute and time-defined assessment than does collection of urine. Use of saliva-based cortisol measures offers researchers the opportunity to assess this "stress hormone" without the reactivity, practical restraints, and ethical problems inherent in more invasive blood or urine sampling procedures (Kirschbaum & Hellhammer, 1989).

One problem associated with salivary measures in stress research has been concern about flow rate. Among its other effects, stress causes a reduction in salivary flow, which can alter concentrations of substances found in saliva. However, cortisol is small and highly lipid-soluble; thus it can diffuse through cell membranes and into saliva. Consequently, saliva flow has little or no effect on salivary cortisol levels (e.g., Kirschbaum & Hellhammer, 1989; Vining & McGinley, 1984). This allows the researcher to measure salivary cortisol without concern about the changes in the rate or composition of saliva due to SNS arousal. Correlations between salivary and blood measures of cortisol suggest that they are accurate reflections of each other, frequently reaching or exceeding .9 (80% of total variance shared; Kirschbaum & Hellhammer, 1989). Temporal relationships also appear to be strong. In one study, salivary cortisol levels increased within 1 minute after 5-milligram injections of cortisol, suggesting that there is rapid transfer of cortisol from blood to saliva (Walker, Raid-Fahmy, & Read, 1984). Peak salivary levels in this study trailed blood peak plasma levels by only 1 or 2 minutes. The rate of disappearance of salivary cortisol also appears to be comparable to that of plasma cortisol, and the same circadian rhythms appear to be reflected in plasma and saliva. However, absolute levels of cortisol in these two media differ. Although plasma and saliva levels are highly correlated, salivary measures show concentrations up to 50 percent lower than and may vary considerably from plasma samples (Kirschbaum & Hellhammer, 1989).

Salivary cortisol has been shown to increase with exposure to physical and psychological stressors (e.g., Bassett, Marshall, & Spilane, 1987; Stahl & Dorner, 1982). The value of a nonreactive method of measurement was highlighted in studies of stress and cortisol levels as a function of public speaking. An increase in salivary cortisol was observed when subjects were asked to give a 15-minute speech, but because samples were obtained only just before and after the speech, time course issues could not be explored. Because saliva collection is not as reactive as is blood sampling, Lehnert et al. (1989) were able to collect additional prespeech samples in a similar study and found that highest cortisol levels were found after the preparation period and just before speaking. This important finding would not have been detectable nor would the study have been done as readily and without concern about reactivity if blood-based measures had been used.

To summarize, salivary measures of cortisol offer a less reactive, less expensive, and more accessible method of measuring stress-related HPA response in humans than do traditional blood and urine measures. Samples can be simply collected by asking subjects to salivate into a plastic tube or container, though some subjects may find this difficult or upsetting. An alternative collection method, using a Salivette (Sarstedt) or similar device, can also be used. The Salivette incorporates a cotton swab, which subjects chew on to stimulate saliva production and on which saliva collects; a beaker, into which the cotton swab is placed (after 30 to 60 seconds of chewing); and a disposable tube (Hellhammer, Kirschbaum, & Belker, 1987). The beaker is centrifuged, and a small (1-ml) sample of saliva pipetted into the disposable tube (Kirschbaum & Hellhammer, 1989). Samples may be stored at room temperature for more than 2 weeks without appreciable degradation of cortisol levels (Kirschbaum & Hellhammer, 1989).

Assays

Corticosteroids are usually assayed by radioimmunoassay (RIA) techniques that use radioisotopes in immunological processes (e.g., Yalow & Berson, 1971). These assays yield reliable determinations of cortisol in plasma, urine, and saliva (e.g., Walker et al., 1978). Basically, the chemicals of interest (e.g., corticosteroids) are exposed to substances (antisera) that recognize the target chemicals and bind to them. Radioisotopes are used to label these target chemicals (single-antibody RIA) or to compete for binding with the chemicals of interest before quantification (competitive binding RIA). This technique is relatively easy to use, is highly reliable, and has been validated extensively. The RIA techniques currently used have minimal cross-reactivity with other similar biochemical molecules. In other words, modern assays are sensitive and specific, and we are able to measure just what we want to measure. Older techniques (e.g., spectrophometric) that were used before the mid-1970s were not as specific and may have included measurement of related chemicals as well as target CO compounds.

Considerations for Use of Corticosteroid Measures

The primary consideration for use of CO measures in stress studies is whether cortisol is a likely mechanism or point of interaction with other systems and/or their effects. Studies of stress and immune system change might consider cortisol as a possible mechanism by which immunosuppression might occur. In such a case, measures of cortisol or CO metabolites could serve a useful function. Similarly, in studies of stress and eating or stress and metabolic changes, measures of cortisol may be desirable or even necessary. If one is interested in the effects of stress on synthesis of epinephrine, changes in diurnal rhythms, mobilization of fat-derived energy, or any of the other outcomes in which cortisol may be involved, measures of CO are important. The costs are far less than for catecholamines and can be as low as $10 per sample in one's own lab. Finally, logistic issues may indicate the use of cortisol or corticosterone when hormonal measures are needed, but prevailing conditions make the study of other endocrines difficult. For example, cortisol is an excellent measure to verify stress when other variables (e.g., movement, collection procedures, timing of samples) would invalidate catecholamine measures.

There are fewer problems or limitations on the use of cortisol measures than of catecholamine measures. Cortisol is neither very sensitive to movement and effort nor immediately responsive and unstable. Consequently, blood-sampled measures of cortisol reflect more stable arousal and can be used to represent an hour or more of mood and experience. However, its effects seem somewhat limited and difficult to detect as compared to those of catecholamines in human subjects, and the relative sensitivity of the HPA system may result in smaller changes and less overall variance explained. Research evidence suggests that psychological variables including stress and conditions such as novelty, unpredictability, and uncontrollability are associated with increases in HPA activity and cortisol release (e.g., Mason, 1975b). However, in many of these studies, a substantial percentage of subjects do not respond (i.e., do not show elevated cortisol levels) after exposure to a stressor, possibly owing to the strong effects of novelty, emotional involvement, and suspenseful anticipation of noxious events (Kirschbaum & Hellhammer, 1989). The sensitivity of the HPA to a variety of different conditions and events may make it less useful as a specific indicator of stress, and the interpretive issues raised by nonresponders or even people who show inverted responses have made cortisol measures more difficult to interpret.

Other Hormonal Measures

Endogenous Opioid Peptides

Depending on the focus of research, a number of other endocrine changes associated with stress may be studied. Endogenous opioid peptides (EOPs), such as beta-endorphin, leu-enkephalin, met-enkephalin, and dynorphin, have been studied as determinants or mechanisms in stress–drug interactions, as factors in eating or smoking, as mechanisms of stress-induced immune system changes, and in a number of other roles. EOPs are present in small amounts in circulation, but their functions in the periphery are not clear. However, there is evidence that EOPs increase during stress and may play a role in the cognitive or emotional after-effects of stress (e.g., Davidson, Hagmann, & Baum, 1991). In human studies, EOPs might be measured if there was reason to believe they had some specific effects that need to be examined (e.g., in studies of stress-related changes in natural killer cell cytotoxicity, see Kiecolt-Glaser & Glaser, Chapter 10, this volume). However, for most purposes, the small amount found in circulation and the fact that the significance of peripheral EOPs is not clear suggest that using these measures in human research is not ordinarily relevant. There may be some exceptions to this generalization because some opioid-acting peptides ("exorphins") have been found in milk and in plant proteins. These chemicals have received little study compared with the better known EOPs.

Measurement of EOPs can be extremely important in clinical studies in which cerebral spinal fluid samples are available, or in animal research in which brain tissue can be collected and analyzed. Animal research has revealed three basic EOP classes: proopiomelanocortin (POMC) peptides, enkephalin pentapeptides, and prodynorphin peptides. Behavioral analyses suggest that these peptides are associated with pain relief and may moderate effects of addictive drugs, eating behavior, and stress. Assay of EOPs is ordinarily done with RIA techniques. However, molecular assays of messenger RNA (mRNA) for specific EOP classes and opioid affinity binding are more popular and widely used than in the past. Currently, these techniques require specialized training, but advances in technology in this area are so staggering that one may find molecular biological assays as user friendly in a few years as are RIA techniques today.

Serotonin

Serotonin (5-HT) is an indolealkyl amine or indoleamine that is found in chromaffin cells of the intestinal mucosa, in the bloodstream (e.g., in platelets), in the central nervous system, and in the brain (primarily in the pineal body). Although the central effects of serotonin are profound, no more than 2 percent of the 5-HT in the entire body is actually in the brain. Among its roles are actions in muscle contraction, feeding, sleep, mood, metabolism, and neuromodulation. Recent research has linked serotonergic activity with obsessive-compulsive disorder, depression, premenstrual syndrome, and eating disorders (e.g., Bastani, Arora, & Meltzer, 1991; Delgado et al., 1991; McBride, Anderson, Khait, Sunday, & Halmi, 1991; Price, Charney, Delgado, & Henninger, 1991; Rapkin, Reading, Woo, & Goldman, 1991; Walsh, Ware, & Cowen, 1991). Because 5-HT is present in blood platelets, some researchers have suggested that blood platelets can be used as models of serotonergic activity in nerve terminals (e.g., Rothman, 1983). Because platelets are readily accessible, studies of serotonin can focus on this important chemical directly without relying on indirect assays of metabolites or assays of excreted unmetabolized 5-HT.

Serotonin is usually assayed by radioenzymatic procedures. Conceptually, this assay is similar to the REA for catecholamines except that the two enzymes used are N-acetyltransferase and hydroxyindole-O-methyltransferase and the labeled metabolite is [3H] melatonin. The assay is sensitive and reliable but requires experienced technicians and the use of radioisotopes. Serotonin also can be assayed by HPLC.

Other Hormones

In addition to these chemicals, prolactin, thyroid hormones, growth factor, insulin, gonadal hormones, and other peptides show changes associated with stress. The meaning of changes in these compounds is not always clear, and in some cases research is not even clear on the direction of change one should expect to see. In some cases, it is useful to use one or another of these measures, as in studies showing a role for thyroid hormones in posttraumatic stress disorder (PTSD) (e.g., Mason, Kosten, Southwick, & Giller, 1990) or in studies of stress and immune system changes (where prolactin or growth hormone may have an effect). Further, most of these measures must be made from blood, limiting the nature of human studies that can be done using them. RIAs are the most common assay techniques for these chemicals today. As mentioned previously, this approach is user friendly, but requires skilled technicians and the use of radioisotopes.

Conclusions and Practical Considerations

We have not intended to give the impression that adrenal and pituitary hormones are the only hormones involved in stress, nor do we intend to suggest that they are more important than are others. Ultimately, the relative usefulness of any measure depends on the question one is asking, and it is no different when endocrine measures are used. However, adrenal hormones, particularly the catecholamines and glucocorticoids, are clearly established as principal drivers of stress and are reliable indices of stress. It is also likely that E, NE, and cortisol or corticosterone are the primary mechanisms by which stress affects other systems, supplementing, extending, and broadening the responses generated by neural stimulation. Regardless, measures of catecholamines and glucocorticoids are the most common endocrine markers in stress research, are the best understood, and in some ways, the easiest to use. With advances in measurement and understanding of other compounds such as serotonin, we suspect that they will be increasingly recognized as critical chemicals in stress. Regardless, catecholamines and cortisol still lead in obvious relevance.

Rationale for Using Hormonal Measures

The basis for using hormonal measures in stress research is the observation that most systems in the body show changes during stress and that hormonal markers of these changes are reliably associated with stress. Conceptually, the central role of SNS and HPA activation in stress provides ample justification for measuring hormonal changes. Because it is basic to arousal and related bodily changes, endocrine activity may be a better marker of physiological changes during stress than are more readily obtainable measures such as blood pressure, heart rate, finger temperature, or palmar sweating. In addition, measures of endocrine changes during stress allow examination of SNS and HPA activation as mechanisms of other aspects of stress arousal or of stress effects on disease processes.

What Questions Can an Investigator Answer?

On a very basic level, any of these measures, with appropriate qualifications, can be used as a check to see if an individual is experiencing stress. If one is manipulating stress by exposing participants to acute laboratory stressors such as a cold pressor task, noise exposure, or mental arithmetic, blood-derived measures of catecholamine or cortisol levels, if appropriately timed, can provide a good manipulation check. Or, if one is trying to determine whether stress and associated arousal are experienced in a given situation, as in studies of natural and human-made disasters (e.g., Baum, Gatchel, & Schaeffer, 1983; Schaeffer & Baum, 1984) or in studies of isolated and confined groups (Carrere, 1983; Nespor, Suedfeld, Acri, & Grunberg, 1993), these measures are useful as well. They are particularly valuable when used as one of several simultaneously measured indices of stress.

In general, multilateral assessment of stress is useful when, for example, behavioral and endocrine data converge (Baum, Grunberg, & Singer, 1982). However, when responses decouple and are not highly correlated with each other, interpretation is more difficult; if a subject reports no distress but exhibits elevated catecholamines or cortisol, the situation can be explained in several different ways. It could reflect denial as in a participant who denies or represses experienced distress while experiencing arousal and other bodily changes associated with stress. Alternatively, arousal could be due to some other variable, such as experiences immediately prior to assessment, activity level and exercise, diet, drug use, or other extraneous factors. Or, in chronic stress situations, elevated hormone levels could reflect new baseline levels and long-term enhancement of endocrine activity; such levels would therefore be viewed as resting levels experienced in the absence of stress. Presumably, in such a situation, application of new stressors would provoke still higher levels of endocrine activity (e.g., Fleming et al., 1987).

As we have suggested previously, using endocrine measures in stress research also provides the opportunity to assess and study mechanisms of change in other systems or of stress-induced pathogenesis. Adrenal hormones are putative mediators of stress-related immune system change, and measurement of E, NE, and/or cortisol permits analysis of the role of baseline levels or short-term changes in these adrenal hormones in acute immunity–stress interactions or more chronic changes in persistent stress situations (e.g., Manuck, Cohen, Rabin, Muldoon, & Bachen, 1991; McKinnon, Weisse, Reynolds, Bowles, & Baum, 1989; Zakowski, Hall, & Baum, 1992; Zakowski, McAllister, Deal, & Baum, 1992). In addition, measuring these hormones may permit examination of endocrine bases of behavioral influences in heart disease, hypertension, cancer, and other diseases that might be affected by stress.

Choosing an Appropriate Measure

The most confusing aspect of using endocrine measures is the fact that there are so many important endocrine changes associated with stress. In addition, most may be assayed in blood or urine, and some can be measured in saliva. How one goes about reviewing these possibilities and selecting the best measures is basic to the design of good studies and contributes heavily to the success or failure of these studies. Ultimately, the most important determinant of the value of a particular measure is the question being addressed. In addition, the time frame of interest and the logistics of collecting, preserving, and assaying samples are important considerations in selecting a measure. Because there are so many possibilities, the decision seems more difficult than it really is. Careful consideration of these issues will usually produce a clear choice for the hormone and type of measurement best suited to one's needs.

Relative Value

All of the hormones we have discussed have important roles in bodily function, but some are more reliable or more readily interpretable. If one needs to verify that a subject is experiencing stress, adrenal hormones provide a suitable check because they show reliable increases during stress and because they are relatively easy to measure. As we have suggested several times, they also have broad impact, affecting most systems in the body and reflecting basic arousing properties of bodily function. These hormones are likely to be involved in systemic changes throughout the body. Of these measures, choices can be made on the basis of the different half-lives and responsiveness of E, NE, and cortisol (see earlier discussion of stability and turnover). However, for manipulation checks, these measures may be unnecessarily costly and require more invasive and reactive procedures than are required. Adding endocrine measures to a study adds considerable procedural complexity, and one would ordinarily not use them unless other outcomes and relationships were of interest. For example, age and gender appear to have effects on release and/or clearance of catecholamines and the elasticity of the HPA axis; the effects of these influences will be different depending on whether urine or blood samples are used.

Logistical Issues

There are some logistical issues that should be considered before one uses biochemical measures in stress research (Grunberg & Singer, 1990). First, the appropriate assays and measurement techniques (e.g., blood, urine) must be determined. This decision must be made in advance because the particular assay to be used (e.g., REA vs. HPLC vs. RIA) requires different sample handling and preservation. Once the decision is made regarding the specific techniques to be used, plans must be made for sample collection, processing, storage, and transport to the assay laboratory. Once samples are collected, immediate centrifugation may be necessary, but sometimes the centrifuge is not readily accessible to the experimental site. Once samples are appropriately collected and processed (with chemical preservative treatment as well as centrifugation), they must be stored in freezers at $-20°$ or $-70°C$ and maintained until the assays are performed. They should be assayed together to control for interassay error, and controls for assay variability should be included as well. These kinds of logistical issues are not insurmountable, but they should be addressed during the planning stage of a study to avoid needless errors and problems.

Final Note

We have described the value, and issues involved in the use, of neuroendocrine measures in stress research and have conveyed some of the usefulness of such investigations. To some extent, however, this chapter reflects the transitory state of the art, since advances in measurement and detection occur quite rapidly. For example, studies of platelets suggest that measurement of catecholamines in this red blood cell may reflect catecholamine activity over a period of several days (Rothman, 1983). If true, platelet catecholamines would constitute a useful measure of chronic stress and aggregate release over a period of days. Similarly, new studies of receptor densities and receptor binding on lymphocytes hold great promise for understanding the dynamics of stress-related hormonal activity. We recognize that this approach is peripheral to the activities and experience of many stress researchers, but we believe that the history of the field demonstrates the importance of these measures. In our own research experience with human and animal subjects, we have found these measures to be valuable despite the many questions and problems that this conclusion has forced us to face. Systematic incorporation of neuroendocrine measures in research can provide a valuable tool for probing a basic level of stress response and ultimately will help us solve the mysteries of the human organism under duress.

References

Bassett, J. R., Marshall, P. M., & Spilane, R. (1987). The physiological measurement of acute stress (public speaking) in bank employees. *Psychophysiology, 5,* 265–273.

Bastani, B., Arora, R. C., & Meltzer, H. Y. (1991). Serotonin uptake and imipramine binding in the blood platelets of obsessive-compulsive disorder patients. *Biological Psychiatry, 30,* 131–139.

Baum, A. (1990). Stress, intrusive imagery, and chronic distress. *Health Psychology, 2,* 653–675.

Baum, A., Gatchel, R. J., & Schaeffer, M. A. (1983). Emotional, behavioral, and physiological effects of chronic stress at Three Mile Island. *Journal of Consulting and Clinical Psychology, 51,* 565–572.

Baum, A., Grunberg, N. E., & Singer, J. E. (1982). The use of psychological and neuroendocrinological measurements in the study of stress. *Health Psychology, 1,* 217–236.

Berne, R. M., & Levy, M. N. (1983). *Physiology.* St. Louis: Mosby, 290, pp. 1033–1068.

Cannon, W. B. (1914). The emergent function of the adrenal medulla in pain and the major emotions. *American Journal of Physiology, 33,* 356–372.

Carrere, S. (1993). *Physiological and Psychological patterns of acute and chronic stress during winter isolation in Antarctica.* Unpublished doctoral dissertation. University of California, Irvine, CA.

Davidson, L. M., Hagmann, J., & Baum, A. (1991). An exploration of a possible physiological explanation for stressor aftereffects. *Journal of Applied Social Psychology, 20,* 869–880.

Delgado, P. L., Price, L. H., Miller, H. L., & Salomon, R. M., Licinio, J., Krystal, J., Henninger, G. R., & Charney, D. S. (1991). Rapid serotonin depletion as a provocative challenge test for patients with major depression: Relevance to antidepressant action and the neurobiology of depression. *Psychopharmacology Bulletin, 27,* 321–330.

Durrett, L., & Zeigler, M. (1980). A sensitive radioenzymatic assay for catechol drugs. *Journal of Neuroscience Research, 5,* 587–598.

Fleming, I., Baum, A., & Weiss, L. (1987). Social density and perceived control as mediators of crowding stress in high-density residential neighborhoods. *Journal of Personality and Social Psychology, 52*(5), 899–906.

Grunberg, N. E., & Singer, J. E. (1990). Biochemical measurement. In J. T. Cacioppo & L. G. Tassinary (Eds.), *Principles of Psychophysiology: Physical, social and inferential elements* (pp. 149–176). New York: Cambridge University Press.

Hellhammer, D. H., Kirschbaum, C., & Belken, L. (1987). Measurement of salivary cortisol under psychological stimulation. In J. N. Hingtgen, D. H. Hellhammer, & D. Huppmann (Eds.), *Advanced methods in psychobiology.* (pp. 281–289). Toronto: Hogrege.

Kirschbaum, C., & Hellhammer, D. H. (1989). Salivary cortisol in psychobiological research: An overview. *Neuropsychobiology, 22,* 150–169.

Lehnert, H., Beyer, J., Walger, P., et al. (1989). Salivary cortisol in normal men. In Weiner, I. Florin, & D. H. Hellhammer (Eds.), *Frontiers in stress research* (pp. 392–394). Toronto: Huber.

Manuck, S. B., Cohen, S., Rabin, B. S., Muldoon, M. F., & Bachen, E. A. (1991). Individual differences in cellular immune response to stress. *Psychological Science, 2,* 111–115.

Mason, J. W. (1975a). A historical view of the stress field. *Journal of Human Stress, 1,* 22–36.

Mason, J. W. (1975b). Emotion as reflected in patterns of endocrine integration. In L. Levi (Ed.), *Emotions: Their parameters and measurement* (pp. 143–182). New York: Raven Press.

Mason, J. W., Kosten, T. R., Southwick, S. M., & Giller, E. L. (1990). The use of psychoendocrine strategies in post-traumatic stress disorder. *Journal of Applied Social Psychology, 20,* 1822–1846.

McBride, P. A., Anderson, G. M., Khait, V. D., Sunday, S. R., & Halmi, C. (1991). Serotonergic responsivity in eating disorders. *Psychopharmacology Bulletin, 27,* 365–372.

McKinnon, W., Weisse, C. S., Reynolds, C. P., Bowles, C. A., & Baum, A. (1989). Chronic stress, leukocyte subpopulations, and humoral response to latent viruses. *Health Psychology, 8,* 389–402.

Nespor, S. M., Suedfeld, P., Acri, J. B., & Grunberg, N. E. (1993). *Reactive irritability and stress in an isolated environment.* Presented at the annual meeting of the American Psychological Association, Toronto, Canada.

Price, L. H., Charney, D. S., Delgado, P. L., & Heninger, G. R. (1991). Serotonin function and depression: Neuroendocrine and mood responses to intravenous *l*-tryptophan in depressed patients and healthy comparison subjects. *American Journal of Psychiatry, 148,* 1518–1525.

Rothman, A. (1983). Blood platelets in psychopharmacological research. *Journal of Progress in Neuropsychopharmacology and Biological Psychology, 7,* 135–151.

Schaeffer, M. A., & Baum, A. (1984). Adrenal cortical response to stress at Three Mile Island. *Psychosomatic Medicine, 46,* 227–237.

Selye, H. (1976). *The stress of life.* New York: McGraw-Hill.

Stahl, F., & Dorner, G. (1982). Responses of salivary cortisol levels to situations. *Endocrinology, 80,* 158–162.

Vining, R. F., & McGinley, R. A., (1984). Transport of steroid from blood to saliva. In G. F. Read, D. Riad-Fahmy, & R. F. Walker (Eds.), *Radioimmunoassays of steroids in saliva* (pp. 56–63). Cardiff: Alpha Omega.

Walker, R. F., Raid-Fahmy, D., & Read, G. F., (1978). Adrenal status assessed by direct radioimmunoassay of cortisol in whole saliva or parotid saliva. *Clinical Chemistry, 24,* 1460–1463.

Yalow, R. S., & Berson, S. A. (1971). In W. D. Odell & W. H. Daughaday (Eds.), *Principles of competitive protein-binding assays.* Philadelphia: J. B. Lippincott.

Zakowski, S. G., Hall, M. H., & Baum, A. (1992a). Stress, stress management, and the immune system. *Applied and Preventive Psychology, 13,* 1–13.

Zakowski, S. G., McAllister, C. G., Deal, M., & Baum, A. (1992b). Stress, reactivity, and immune function in healthy men. *Health Psychology, 11,* 223–232.

9

Measurement of Cardiovascular Responses

David S. Krantz and Jennifer J. Falconer

Introduction and Historical Background

The cardiovascular system is highly responsive to a variety of psychological and behavioral states. This observation was first noted by ancient Egyptian physicians and philosophers who believed that the heart was the seat of emotion (Hassett, 1978), and promoted by many generations of romantic writers who observed cardiac changes associated with intense emotions such as love, fear, joy, and sadness (Dunbar, 1947). Only in the seventeenth century was the primary function of the heart of pumping blood throughout the body discovered by William Harvey (Willis, 1965). With the advent of techniques for accurate measurement of various cardiovascular functions, modern investigators have noted that blood pressure and heart rate are usually low during restful situations and sleep, increase during periods of wakefulness and moderate activity, and tend to become elevated during periods of intense activity and stress (e.g., Gunn, Wolf, Block, & Person, 1972; Pickering, Harshfield, Kleinert, Blank, & Laragh, 1982); Schneiderman, Ironson, & McCabe, 1987). Given the intimate involvement of the cardiovascular system in processes such as emotion and arousal, and the increasing attention devoted to the effects of acute and chronic stress in the development of cardiovascular disorders such as hypertension and coronary artery disease, it is not surprising that cardiovascular variables are important and widely used measurement tools employed in modern stress research (Krantz, Baum, & Singer, 1983; Matthews et al., 1987; Schneiderman, Weiss, & Kaufman, 1989).

This chapter provides a general overview of issues and techniques involved in the selection and utilization of cardiovascular parameters as physiological measures of the stress response, and is directed toward students and investigators who are considering the use of such measures in research. More comprehensive reviews of the physiological basis of cardiovascular psychophysiology and of technical measurement procedures can be found in several recent chapters and volumes (Cacioppo & Tassinary, 1990; Larsen, Schneiderman, & Pasin, 1986; Schneiderman et al., 1989). This chapter begins with a discussion of preliminary considerations involved in the use of cardiovascular measures, followed by a brief overview of the physiological basis of the cardiovascular system and of techniques of cardiovascular measurement. We then consider several additional methodological issues and areas receiving current attention in cardiovascular stress research.

Preliminary Considerations in Interpreting Cardiovascular Stress Measures

The informed use of cardiovascular measures in stress research requires a recognition of complexities at two levels: (1) the physiology of the cardiovascular system, and (2) the psychophysiology of human response to environmental stressors.

Role of the Cardiovascular System in Maintaining Homeostasis

The primary function of the cardiovascular system, which consists of the heart and the blood vessels, is to help maintain adequate blood flow through various bodily tissues in the face of constantly changing metabolic requirements (Papillo & Shapiro, 1990). To accomplish this function of maintaining homeostasis under a variety of physical and psychological conditions, the bodily mechanisms that control blood pressure and the distribution of blood to different tissues of the body involve regulatory systems that interact with one another and are interdependent (see Figure 9.1). For example, cardiac output—the amount of blood pumped by the heart—is influenced by mechanical (e.g., cardiac contractile) factors relating to the stretch of cardiac muscle fibers, by neural influences of the sympathetic branch of the autonomic nervous system, as well as by neurohumoral factors, such as the circulating hormones epinephrine and norepinephrine. The peripheral vasculature is also influenced by neurohumoral and mechanical processes. Thus, adjustments in the energy needs of bodily tissues (e.g., during exercise or psychological stress) result in a complex pattern of cardiovascular adjustments involving neural, endocrine, and mechanical factors. Changes in any one component of the system necessarily affect other components of the system (Forsyth, 1974). For example, when blood pressure increases abruptly, a strong response (the so-called baroreceptor reflex) is evoked which involves feedback from cardiovascular control centers of the brain and the autonomic nervous system, immediately to reduce heart rate and the force of the heart's contraction and to relax blood vessels in order to return the blood pressure rapidly to its original level. The opposite response occurs in situations where blood pressure is abruptly reduced. Thus, in situations where the baroreceptor reflex is induced, heart rate tends to decrease as blood pressure increases, and an inverse relationship is observed (Papillo & Shapiro, 1990). This suggests that measuring a single physiological index of heart rate or blood pressure as an invariant marker of "stress" in such situations may yield equivocal information. Instead, because cardiovascular adjustments consist of an integrated pattern of responses, it is important to select judiciously a sufficient number of response measures to allow for the response pattern and its variation to be identified (Schneiderman & Pickering, 1987).

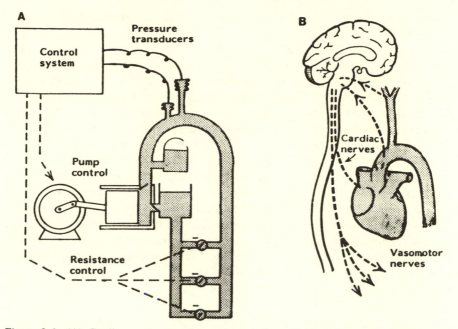

Figure 9.1. (A) Cardiovascular function and control are illustrated by a simple hydraulic model. A pulsatile pump propels fluid through a distribution system with a high driving pressure maintained by a feedback control system with input from pressure sensors and output to both the pump and the valves controlling outflow. (B) The human circulatory system has corresponding features and similar functions. (From *Cardiovascular Dynamics*, 4th edition, by R. F. Rushmer. Copyright © 1976 by W. B. Saunders. Reprinted with permission.)

Situational Factors and Patterns of Cardiovascular Responses

With regard to the interplay of psychological and physiological triggers of the cardiovascular response to stress, investigators have noted that cardiovascular responses to stress are not invariably part of a "nonspecific" or generalized stress response (see Cohen, Kessler, & Gordon, Chapter 1, this volume). Instead, tasks or situations with different psychological or behavioral demands produce stereotypically different patterns of physiological responses (termed *situational stereotypy* (Lacey, 1967) or *situational specificity* (Engel, 1972). Thus, for example, stressful or arousing tasks that require active efforts to cope may produce a different pattern of cardiovascular responses (increased heart rate and systolic blood pressure) in comparison to tasks that require quiet attentiveness to the environment or vigilance (producing momentarily decreased heart rate, decreased cardiac output, and reduced blood flow to skeletal muscle. Krantz, Manuck, & Wing, 1987; Lacey, 1967; Obrist, 1981). The implication of the notion of situational specificity is that different types of stressors can operate on the body via different physiological pathways or have different directional impacts on a particular physiological parameter. Lacey (1967) coined the term *directional fractionation* to refer to dissociations among various measures of physiological and behavioral arousal that often occur in response to stressors.

In addition to specific task demands, the particular emotional reaction elicited by a particular stressor may also determine the observed cardiovascular and neuroendocrine responses (see also Baum & Grunberg, Chapter 8, this volume). Early studies in experimental psychophysiology sought to demonstrate differential heart rate, blood pressure, and neuroendocrine responses during laboratory procedures designed specifically to induce emotions such as fear and anger (Ax, 1953). More recent studies suggest that cardiovascular responses differ across a wide range of negative emotions (e.g., fear, anger, sadness, disgust) and may differ between positive and negative emotions generally (Ekman, Levenson, & Friesen, 1983; Schwartz, Weinberger, & Singer, 1981), although response differences between different emotions are often subtle and difficult to detect. The notion of situational specificity in cardiovascular stress responses suggests that the selection of particular physiological response measures in cardiovascular stress research should be made in the context of an awareness of the stereotyped cardiovascular reactions likely to be produced by the particular stressor under study.

Fundamentals of Cardiovascular Structure and Function

The cardiovascular system consists of (1) the heart, which is comprised of four chambers that function as two pumps in tandem; (2) the vasculature, which consists of vessels that distribute blood (arteries and arterioles) and collect blood (veins and venules); and (3) a system of thin-walled vessels called capillaries. The left-side heart pump (Figure 9.2, left), composed of the left atrium (or left auricle) and the left ventricle, propels oxygen-rich blood through the arterial system to the systemic circulation, onto the target organs where nutrients are removed and wastes added, then through the veins and back into the right atrium. The right-side heart pump, composed of the right atrium and right ventricle, moves the now oxygen-deficient blood through the pulmonary circulation in the lungs, where wastes are removed and oxygen is replaced. Bodily circulation is maintained by a regular cycle of events in the heart (the so-called cardiac cycle): systole, or the contraction of heart muscle during which blood pressure peaks as blood is forced from the heart; and diastole, the relaxation of heart muscle, during which blood pressure reaches its lowest value (Hassett, 1978).

Neural and Hormonal Control of Heart Rate
and Contractile Force

The beating of the heart is an electromechanical event. That is, electrical impulses generated by specialized pacemaker cells within the heart initiate the mechanical contraction of heart muscle or myocardium. The internal cardiac pacemakers, the sinoatrial (SA) and atrioventricular (AV) nodes (Figure 9.2, left) are responsible primarily for controlling heart rate and rhythm of contraction. The orderly transmis-

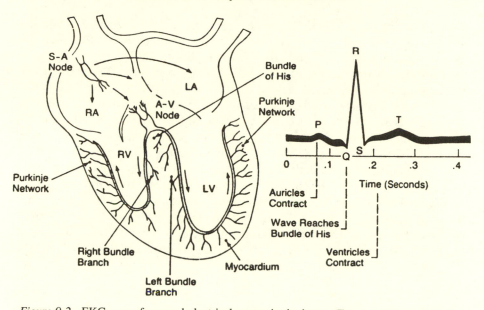

Figure 9.2. EKG wave-form and electrical events in the heart. (From *A Primer of Psychophysiology,* by J. Hassett. Copyright © 1978 by W. H. Freeman & Co. Reprinted with permission.)

sion of electrical activity through the various portions of the heart gives rise to an electrical field—the electrocardiogram or ECG (see Figure 9.2, above)—that can be measured with electrodes placed on the body surface (see later section on measurement). The ECG is a graphical representation of the pattern of electrical activity generated by the heart during each beat, and each portion of the ECG corresponds to a particular portion of the cardiac cycle.

The SA node has an intrinsic rate of firing in the range of 105 to 110 firings per minute, yet in healthy adults at rest, the heart rate is around 70 beats per minute (Papillo & Shapiro, 1990). During sleep, HR may be reduced by 10 to 20 beats per minute, and during emotional stress or exercise, it may exceed 150 beats per minute. These differences in rate result from neural and hormonal influences on the heart, which is innervated by both the sympathetic and parasympathetic branches of the autonomic nervous system.

An increase in the activation of sympathetic nerve fibers innervating the heart causes the SA node to produce an increase in heart rate (also referred to as a positive *chronotropic* effect), and also causes an increase in the force of ventricular contraction (also referred to as a positive *inotropic* effect). When sympathetic activity decreases, the result is decreased heart rate (negative chronotropic effect) and decreased ventricular contraction (negative inotropic effect). In addition to the innervation of the sympathetic cardiac nerves, sympathetic activation of the heart can also occur hormonally as a result of the release of epinephrine and, to a much lesser extent, norepinephrine into the bloodstream from the adrenal medulla (Larsen et al.,

1986). Sympathetic influences on the heart are mediated by receptors, specifically adrenoreceptors of the beta-1 type. The functions of receptors are discussed in a later section on control of blood flow to the peripheral vasculature.

Parasympathetic fibers travel to the heart via the vagus nerve. The vagus causes heart rate deceleration and decreased force of cardiac muscle contraction when vagal activity increases, and heart rate acceleration when parasympathetic activity decreases. Thus, under various conditions, sympathetic and parasympathetic influences on the heart may operate either synergistically or antagonistically to alter heart rate. Because a heart rate increase may result either from sympathetic activation or parasympathetic withdrawal, measures of heart rate alone may provide ambiguous information concerning the precise mechanism(s) of neural influences on the heart (Papillo & Shapiro, 1990). The investigator interested in accurately inferring neurogenic changes that underlie heart rate alterations must therefore assess additional, more sophisticated measures of cardiac function (e.g., analysis of heart rate variability, assessments of left ventricle performance, etc.) (see Larsen et al., 1986; Wilson, Lovallo, & Pincomb, 1989).

Control of Cardiac Output

The rate at which blood is pumped from the left ventricle of the heart is known as the cardiac output (CO). CO can be determined as the product of the heart rate and the volume of blood ejected by the heart during each beat (stroke volume). Stroke volume is determined by several mechanical factors including (1) the volume of blood in the ventricle prior to its contraction, (2) the resistance to the output of the heart during each cardiac cycle imposed by the pressure in the arterial circulatory system, and (3) the strength of contraction (or contractility) of the ventricle. Contractility of the ventricle is increased when more blood is present in the ventricle prior to contraction. More precontraction filling causes a greater stretch of ventricular muscle and stronger subsequent ventricular contraction (i.e., a positive inotropic effect). The relationship between the filling of the ventricle and contractility is known as "Starling's law of the heart" (Patterson, Piper, & Starling, 1914).

In addition to the three aforementioned mechanical factors influencing cardiac contractility, the sympathetic nervous system can also influence the force of ventricular contraction, as described in an earlier section of this chapter. Figure 9.3 illustrates the multiple interacting neural and mechanical influences on cardiac output. Cardiac output and the total resistance to flow provided by the blood vessels, in turn, are important determinants of blood pressure (see below).

Control of Blood Flow to the Peripheral Vasculature

Large and small vessels, termed arteries and arterioles, respectively, serve as conduits to carry the blood from the heart to the various tissues in the body. Arteries have thick walls containing smooth muscle and elastic fibers. When blood is ejected from the ventricles with each beat of the heart, the arteries passively expand, owing

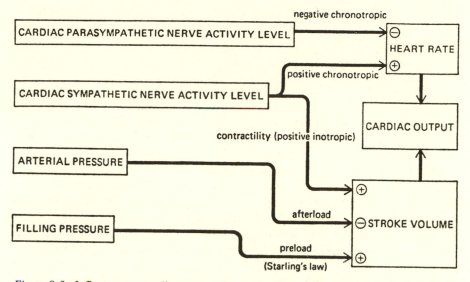

Figure 9.3. Influences on cardiac output. (From *Cardiovascular Physiology*, by L. J. Heller & D. E. Mohrman. Copyright © 1981 by McGraw-Hill. Reprinted with permission.)

to their elastic nature. The arteries subsequently recoil, thereby causing blood to be propelled to adjacent vascular segments (Papillo & Shapiro, 1990). With increasing distance from the heart, the inside diameter of the arteries decrease and these arterioles contain a higher proportion of smooth muscle and are less elastic than larger arteries. In contrast to the passive expansion of arteries, the diameter of arterioles is controlled primarily by neural innervation from the sympathetic nervous system. This sympathetic stimulation usually produces a moderate constriction of the blood vessels, and variations in the degree of vasoconstriction or vasodilation of these vessels are caused primarily by changes in activity of the sympathetic nervous system (Papillo & Shapiro, 1990).

The action of the sympathetic nervous system on the small arteries and arterioles in the various organs of the body occurs via increases or decreases in the stimulation of *receptors* in smooth muscle cells in arteries. Two important types of adrenergic (or sympathetic) receptors are *alpha-* and *beta-* receptors, and an increase in sympathetic stimulation may cause *either* constriction *or* dilation in a particular area of the vasculature (vascular bed), depending on the functions of receptors in that vascular bed. For example, stimulation of alpha-receptors in blood vessels serving the skin, skeletal muscles, heart, and visceral organs (e.g., the gastrointestinal system) causes vasoconstriction, and sympathetic stimulation of beta-receptors in the heart and skeletal muscles causes vasodilation. More detailed discussion of various subtypes of adrenergic receptors in the cardiovascular system can be found in Larsen et al. (1986) and Weiner (1980). Suffice it to say here that research suggests that the net effect on systemic blood flow of the stress ("fight or flight" response) is a shifting of blood away from the skin and visceral organs toward the skeletal muscles. This is accomplished by vasoconstriction and decreased blood flow in the skin and visceral organs (e.g., gastrointestinal system, kidney), and vasodilation with increased blood flow in the skeletal muscle (Brod, Fencl, Heil, & Jirka, 1959; Williams, 1987).

Blood Pressure

A pressure force is needed in order to drive the output of the heart through the various components of the circulatory system and to regulate the blood flow to various organs of the body. Systemic arterial blood pressure (BP) refers to the force exerted by the blood against the walls of the blood vessels. Blood pressure is determined by the product of the cardiac output (CO) and the total resistance to blood flow in the systemic circulation, or total peripheral resistance (TPR); this relationship is expressed by the equation: arterial BP = CO × TPR. Thus, factors that affect either CO or TPR, without producing a compensating change in the other parameter, will affect BP. It also follows that any process or intervention that affects BP must do so by altering CO, TPR, or both of these (Papillo & Shapiro, 1990).

It is possible to describe changes in arterial pressure throughout the various points in the cardiac cycle with an arterial pressure curve (see top portion of Figure 9.4). Blood pressure peaks during systole at the point of ventricular contraction, and reaches its minimum during diastole when the cardiac muscle relaxes prior to its next beat. Blood pressure is usually expressed as the systolic over the diastolic value, with the systolic value necessarily being the higher of the two. The typical units of measurement for BP are millimeters of mercury (or mmHg). There is considerable variability in blood pressure among individuals, but the "normal" blood pressure value for young adults under resting conditions is in the range of 120/80 mmHg. Two terms that are also frequently used are (1) *mean arterial pressure*, which refers to the average pressure during the cardiac cycle and reflects the average effective pressure that drives the blood through the circulatory system; and (2) *pulse pressure*, which refers to the systolic pressure minus the diastolic pressure.

Cautions Regarding Inferences Drawn from Measurement of BP

Factors such as gender, race, weight, age, health status, and consumption of salt, caffeine, nicotine, and so on, as well as exposure to exercise or psychological stress can all influence blood pressure levels (Matthews et al., 1987; Papillo & Shapiro, 1990). For these reasons, a single blood pressure measurement can be highly unreliable, and multiple measures of resting BP should be taken. In addition, when BP measures are used to diagnose the clinical condition of high blood pressure, BP should be taken on multiple occasions as well.

As noted above, blood pressure regulation involves the complex interaction of many physiological processes. Blood pressure is maintained in a narrow range and plays a role in controlling the function of the cardiovascular system through the operation of various negative feedback mechanisms that operate on both short-term (e.g., the baroreceptor reflex) and long-term (e.g., fluid retention and release by the kidneys) bases. As a result of these complex regulatory mechanisms, similar changes in blood pressure produced by two different situations or in the same individual at two points in time may be produced by completely different patterns of cardiovascular changes. As a result, the measurement of blood pressure provides only a very general—although important—index of cardiovascular activity, and should therefore be thought of as indexing multiple physiological processes, rather than any single physiological mechanism. To be able to specify precisely the mechanisms resulting in a particular change in blood pressure, it is necessary to assess measures of the various factors that determine blood pressure response (e.g., cardiac output, total peripheral resistance, and/or sympathetic influences on the heart). Techniques for assessing some of these processes (e.g., forearm vascular resistance) are described briefly later in this chapter. However, a detailed discussion of assessment of these various factors is beyond the scope of this chapter, and the reader is referred to excellent treatments of this issue in Larsen et al. (1986), Papillo and Shapiro (1990), and Schneiderman et al. (1989).

Noninvasive Measurement of Cardiac Function

Although numerous cardiovascular functions (e.g., sensitive and specific indices of cardiac contractility, cardiac output, and sympathetic and parasympathetic influences on the heart, etc.) can be measured without the use of highly invasive measurement techniques, the measurement of three cardiovascular parameters that are most commonly and easily utilized in stress research—heart rate, blood pressure, and peripheral vasoconstriction—are considered in this chapter. Comprehensive reviews of techniques for noninvasive assessment of a variety of more sophisticated cardiovascular parameters can be found in Cacioppo and Tassinary (1990), Coles, Donchin, and Porges (1986), and Schneiderman et al. (1989).

Blood Pressure Measurement

The "gold standard" for the assessment of blood pressure is the direct recording of intra-arterial blood pressure, which is invasive and also involves a degree of risk to the subject. In terms of noninvasive measurement techniques, the *auscultatory* method to estimate an individual's systolic and diastolic blood pressure is still the most widely used both in clinical practice to measure blood pressure, and in clinical research where automated BP recorders may be used (Pickering & Blank, 1989).

The basic principles of the auscultatory technique are illustrated in Figure 9.4. A cuff that inflates is wrapped around the upper arm, and a device, such as a mercury column sphygmomanometer, is attached to measure the pressure in the cuff. The cuff is next inflated to a pressure that is considerably above a normal subject's systolic blood pressure level (e.g., between 175 and 200 mmHg). This cuff pressure around the arm collapses the blood vessels in the upper arm and prevents the blood from flowing into or out of the forearm while the cuff pressure remains higher than

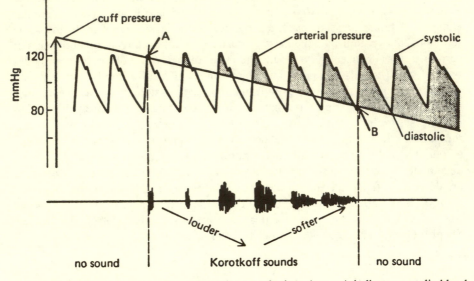

Figure 9.4. Blood pressure measurement by auscultation. Arrow A indicates systolic blood pressure. Arrow B indicates diastolic blood pressure. (From *Cardiovascular Physiology*, by L. J. Heller & D. E. Mohrman. Copyright © 1981 by McGraw-Hill. Reprinted with permission.)

systolic blood pressure. After this inflation, air in the cuff is allowed to withdraw from the cuff so that the cuff pressure falls slowly. As soon as the cuff pressure falls below peak (systolic) arterial pressure, some blood begins to pass through the arteries beneath the cuff when blood pressure is at systolic level. The intermittent flow of blood beneath the cuff produces sounds of turbulence, which can be detected with a stethoscope placed over the radial artery at the inside of the elbow (Heller & Mohrman, 1981). Various sounds of turbulent flow, named *Korotkoff sounds* after the clinician who first noted their presence, are heard when the cuff pressure is between systolic and diastolic arterial pressure. When diastolic pressure is reached, blood flows freely beneath the cuff and no sound is heard. As the cuff is deflated, the highest cuff pressure at which intermittent sounds of turbulent flow are heard is systolic arterial pressure, and the cuff pressure at which the sounds become muffled or disappear is diastolic blood pressure (cf. Heller & Mohrman, 1981).

Sources of Error in Blood Pressure Measurement

Observer error and bias are important sources of error when a conventional sphygmomanometer is used. Since Korotkoff sounds are more discernible and distinct near systolic than near diastolic pressure, measuring diastolic pressure via auscultation requires special care, and is subject to considerable error in untrained individuals. Consistent errors may be produced by differences in auditory acuity, and observers often have a preference for certain blood pressure numbers or digits, recording a disproportionate number of readings ending in 5 or 0 (Pickering, 1968).

The behavioral phenomenon of "white coat hypertension," caused by the presence of a physician, nurse, or other medical professional, is also important. Ayman and Goldshine (1940) and many others subsequently, have observed that blood pressures recorded in the clinic can be considerably higher than pressures taken by the patient at home, using the same technique and with the patient in the same posture. Race and sex of the observer can also influence recorded blood pressure. For example, men tend to have higher pressures when the measurement is taken by a woman than by a man, whereas the opposite is true for a woman (Comstock, 1957; Pickering & Blank, 1989). Many of these sources of error can be minimized by habituating the subject to the experimental situation or setting. Nevertheless, in addition to causing inaccuracies in the estimation of resting blood pressure, the existence of social and behavioral influences on measured blood pressure also provide an illustration of the considerable sensitivity of blood pressure to the effects of social, psychological, and environmental stress. A more detailed discussion of the effects of social context on cardiovascular assessments can be found in Krantz and Ratliff-Crain (1989).

The accurate estimation of blood pressure via the auscultatory method requires an appropriate match between the subject's arm diameter and the size of the BP cuff used for measurement. Thus, in obese subjects, a regular adult cuff size may seriously overestimate BP, and larger cuff sizes must be used. Other special populations and circumstances may also require special consideration in measuring BP accurately. The auscultatory BP assessment technique may underestimate true systolic BP in infants and young children, and a decrease in the distensibility of arteries with aging may result in a tendency toward increased systolic BP without a corresponding diastolic BP increase in elderly subjects (Pickering & Blank, 1989). Because of the effects of postural changes on blood pressure, it is also important to control for posture when assessing BP.

Automated BP Assessment

Various devices and techniques exist to minimize and/or eliminate the observer sources of error in blood pressure assessment. Most widely in current use are automatic and semiautomatic BP recorders (price range between $1500 and $5000 for commercially available monitors). Most of these monitors employ a sphygmomanometer cuff, but may differ in the particular technique used to estimate blood pressure. In addition to the auscultatory technique described above, many of these automated monitors may detect BP via an *oscillometric* method, and thus may differ in their assessments from auscultatory measurements that are based on Korotkoff sound detection. The oscillometric technique estimates BP indirectly based on an algorithm that relies on oscillations of pressure in the cuff during gradual cuff deflation. An advantage of this method is that no microphone transducer need be placed over the brachial artery, so that precise placement of the cuff is no longer a critical factor (Pickering & Blank, 1989). This technique may also be advantageous

in noisy situations where Korotkoff sounds may not be easily heard. The oscillometric technique appears to work reasonably well in most subjects, but may
systematically differ in assessments employing auscultatory measurements. In
choosing from the many available automated monitors, it behooves the investigator
to ask for information regarding the reliability and validity of such devices. A recent
investigation noted that many recorders give readings that consistently differ from
auscultatory assessments by more than 5 mmHg, with no single method of recording being consistently superior (Pickering, Cvetovski, & James, 1986).

Ambulatory Monitoring of Blood Pressure

Since the early 1980s, there have been major advances in the development and
validation of portable automated units that enable researchers to monitor blood
pressure periodically in ambulatory individuals during activities of daily life (see
Harshfield, Hwang, Blank, & Pickering, 1989; Weber & Drayer, 1984). These units
consist of devices capable of automatically inflating conventional arm cuffs and
recording blood pressure at preset intervals during periods when the subject is both
awake and asleep. Price ranges for commonly used ambulatory BP monitors range
from $2500 to $5000. Structured diaries are usually completed by subjects during
ambulatory monitoring periods in order to assess factors such as posture, physical
activity, substance use, and so on, that may affect blood pressure readings.

Ambulatory recording techniques have provided clinicians with the opportunity
to measure blood pressure outside of the clinic setting in order to determine which
individuals are in need of treatment for essential hypertension, and also provide
investigators with a means for examining the effects of a variety of behavioral
factors (e.g., occupational stress, personality, etc.) on blood pressure during a
variety of daily activities. Despite the technological advances in the design of
relatively lightweight and compact monitors, problems inevitably arise during the
use of these devices. For example, care must be taken to ensure that the device is
carefully calibrated against sphygmomanometric readings before measurements are
taken; the cuff does not move from its placement on the subject's arm during the
recording period; that procedures are established for data treatment of lost BP
readings and/or readings that for various reasons appear invalid. As a result of these
issues, a learning period for investigators is often needed before they are able to
utilize automated BP devices effectively. Excellent discussions of methodological
issues in ambulatory blood pressure monitoring, including factors such as reliability
of measurements and data treatment, are found in Harshfield, et al. (1989), Pickering (1993), and Pickering and Blank (1989).

Measurement of Heart Rate

Heart rate and rhythm are among the simplest cardiac functions to measure. Heart
rate can be easily counted by manually palpating the pulse, but automated techniques have the obvious advantages of greater precision, less obtrusiveness to the

subject, and the ability to obtain beat-by-beat information that can be correlated with various behavioral and psychological events. Several automated techniques rely on the visual display of each beat of the heart, and others process heartbeat information and display a number digitally that corresponds to heart rate.

The most commonly used technique for assessing both heart rate and cardiac rhythm is electrocardiography. The electrocardiogram (ECG or EKG) consists of a graphically displayed recording of cardiac electrical activity. It is typically measured in the form of a voltage potential recorded by a set of electrodes on the body surface. The ECG potentials are then written out by a moving pen on paper moving at a constant speed. The electrical activity of the heart, recorded from the body surface as the ECG, is brought about through a repetitive sequence of excitation and recovery of the membranes in the cardiac cells (Wilson et al., 1989). These electrical events precede and initiate the mechanical events of the cardiac cycle (see Figure 9.2).

In a clinical diagnostic setting, the ECG is recorded from as many as 12 different pairs of leads, half of these on the chest and half on the limbs. Each pair of leads detects a difference in electrical potential across the heart, and gives slightly different information about cardiac function (Hassett, 1978). Characteristic positions and names are assigned to each lead. For example, lead I records the potential difference between left arm and right arm, lead II between right arm and left leg, and so on. With a 12-lead ECG, it is possible to detect various irregularities (arrhythmias) in cardiac electrical activity, as well as the presence of other cardiac pathologies such as ischemia (lack of blood flow to cardiac tissue) and myocardial infarction or heart attack.

For applications where heart rate is the only measure of primary interest, the exact positioning of the electrode leads is of little importance because of the robust size of the cardiac bioelectric signal, provided that one electrode is on either side of the heart. A lead II configuration (one lead above the right wrist and the other above the left ankle) is often used because it produces a pronounced complex of QRS waves of the ECG (see Figure 9.2). If the paper speed is known, heart rate can be easily determined as the number of R-waves (usually the highest lead II ECG wave) per 1-minute period. To determine heart rate over a shorter duration of time, one can measure the time between two successive R-waves, known as the R-R interval or heart period, and utilize a formula to transform this information to beats per minute. It is also possible to use a device called a cardiotachometer in conjunction with the ECG. This device measures the length of time between two successive QRS complexes and converts this to a measure of rate in beats per minute. Heart rate assessments can be simply and easily made as well by use of photoplethysmography (see below). Approximate digital readouts of heart rate can also be obtained from the output of most automated blood pressure monitoring devices.

Ambulatory Measurement of the Electrocardiogram

As is the case for blood pressure, reliable techniques have been developed that enable investigators and clinicians to monitor the electrocardiogram continuously as subjects/patients go about their daily activities (see Kligfield, 1989). Such devices consist of lightweight, portable tape recorders issued in conjunction with two or three electrocardiographic leads. The cost of these units ranges between $1500 and $3000. Ambulatory ECG recording has been used to study and/or diagnose clinical phenomena (e.g., cardiac arrythmia, cardiac ischemia) during daily life in patients with known or suspected cardiovascular disease, as well as to examine patterns of heart rate and heart rate variability. Analysis of heart rate variability can be important for the information it conveys regarding parasympathetic influences on the heart as well as for its possible clinical importance in assessing risk of sudden cardiac death (Bigger, Fleiss, Rolnitzky, Steinman, 1993; Jiang et al., 1993). Another particularly promising arena makes use of ambulatory ECG monitoring to link cardiac ischemia (a condition characterized by inadequate blood flow to cardiac tissue, and an important marker of coronary disease) to mental stress (e.g., Krantz et al., 1993).

Measuring Peripheral Blood Flow

Researchers who study the cardiovascular effects of psychological stress and who wish to assess peripheral blood flow commonly use one of two noninvasive techniques. One technique involves the measurement of cutaneous blood flow through the skin capillary beds serving peripheral parts of the body such as the ear or the finger. A second technique assesses the flow of blood to the skeletal muscles, such as the forearm (Papillo & Shapiro, 1990).

Photoplethysmography

Psychophysiological researchers in the 1960s and 1970s commonly used photoplethysmography to assess blood flow to the finger. This measure provides a very rough index of sympathetic activation since the blood vessels in the finger constrict during periods of stress or activation. Photoplethysmography is based on the principle that living tissue and blood have different light-absorbing properties (cf. Papillo & Shapiro, 1990). In one such system, the finger tip is placed between two parts of a transducer consisting of a light source and a photocell (which converts light to electrical energy), and a beam of infrared light is projected toward the photocell. The blood in the finger scatters light in the infrared range, and the amount of light reaching the photocell is inversely related to the amount of blood in the finger. Therefore, when blood vessels in the finger dilate, the increased blood flow allows less light to reach the photocell; when blood vessels constrict, blood flow is lessened and more light reaches the photocell.

Figure 9.5 displays chart tracings taken from a finger photoplethysmograph. Blood flow responses measured via phethysmography can be either *tonic* or *phasic*. Tonic blood flow measurements, which assess longer-term changes made over a relatively extended period of time, are also called blood volume (BV) measurements. They reflect the relative volume of blood remaining in the finger over this time period (Figure 9.5, bottom). Phasic changes (Figure 9.5, top), made on a beat-

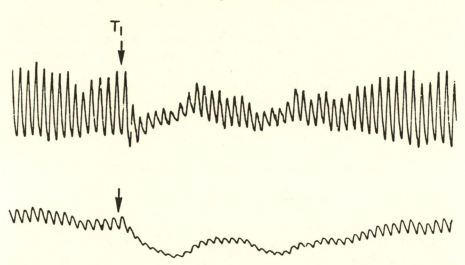

Figure 9.5. Pulse volume (top) and blood volume (bottom) polygraph recordings obtained from output of finger photoplethysmograph. T_1 indicates the onset of vasoconstriction. Pulse volume was recorded using a 1-second time constant. (From *Principles of Psychophysiology: Physical, Social, and Inferential Elements,* by J. F. Papillo & D. Shapiro. Copyright © 1990 by Cambridge University Press. Reprinted with permission.)

to-beat basis, are called pulse volume (PV) measurements and are related to beat-to-beat variations in the force of blood flow (Hassett, 1978). Because PV varies with each beat of the heart, heart rate can be easily quantified from a plethysmograph tracing without necessarily using an ECG.

It is important to note that all of the measure derived from photophlethysmography reflect *relative,* and not absolute changes in blood flow. Use of other techniques, such as venous occlusion plethysmography (see below), is necessary to quantify absolute levels of blood flow. In addition, plethysmograph recordings are highly sensitive to several sources of artifact, including raising or lowering the arm (thereby greatly altering relative blood flow into the finger), and changes in temperature (which cause vasodilation or vasoconstriction).

Venous Occlusion Plethysmography

Of perhaps more current interest is the measurement of skeletal muscle blood flow to the limb. The measurement of forearm blood flow, in conjunction with other indices of cardiovascular function, enables the investigator to assess various components of the "fight-or-flight" response, which involves an increase in skeletal muscle blood flow. To measure blood flow to the forearm, a strain gauge (a device that changes its electrical resistance with stretch) is used in conjunction with two blood pressure cuffs. One BP cuff is placed on the upper arm, and the other is placed around the wrist. The strain gauge is placed around the portion of the forearm having the largest circumference. To measure forearm blood flow, the wrist cuff is inflated to cut off blood flow to the hand, and the upper arm cuff is inflated to a pressure (e.g., 40–50 mmHg) that is above the pressure in the veins but below diastolic arterial pressure. Flow of blood out of the arm via the veins is thereby prevented, but blood flow into the arm via the arteries is unaffected (Larsen et al., 1986; Papillo & Shapiro, 1990). The strain gauge resistance is recorded on a polygraph and/or computer. As the arm becomes engorged with blood, a change in resistance of the strain gauge with increases and decreases in blood flow to the forearm is proportional to the change in the circumference of the strain gauge.

Additional Issues and Current Research Directions

Cardiovascular Reactivity

Investigators in psychosomatic medicine have long been interested in the role of individual differences in cardiovascular and endocrine responses to stress in the development of psychosomatic disorders (e.g., Weiner, 1977). In this regard, considerable recent research has investigated cardiovascular responses to stress as possible markers or contributing factors in the development or progression of coronary heart disease and essential hypertension (Krantz & Manuck, 1984; Matthews et al., 1987). Cardiovascular reactivity is measured in terms of heart rate, blood pressure, or other cardiovascular *changes* in response to stress, as opposed to measuring only resting levels of cardiovascular variables (Krantz & Manuck, 1984; Matthews et al., 1987). An underlying assumption of this research is that by recording changes over resting levels produced by specified laboratory stress conditions, one can obtain an estimate of the individual's responses to the stresses of everyday life. In this regard, the issue of the correspondence between laboratory measures of cardiovascular reactivity and blood pressure and heart rate measures during daily life assessed via ambulatory monitoring is currently being investigated (see Manuck, Kasprowicz, Monroe, Larkin, & Kaplan, 1989).

As noted in an earlier section of this chapter, the concept of stimulus specificity is based on the observation that cardiovascular reactions elicited by behavioral stressors vary in their magnitude and in their patterning, depending on the eliciting situation. In addition, there is considerable evidence that individuals exposed to the same stimulus situations display considerable variability in their cardiovascular responses. Research (e.g., Manuck et al., 1989) suggests that these individual differences are relatively stable over time and are, to some extent, reproducible under varying stimulus conditions.

Choice of an Appropriate Baseline in Cardiovascular Psychophysiology Research

Cardiovascular measures of the stress response are often compared to resting or "basal" levels, and reactive or "stress" levels of cardiovascular function derive much of their meaning by comparison with corresponding measurements made in the same individual under resting conditions (Krantz & Manuck, 1984). Many studies make these control measurements under conditions where the subject is sitting quietly, and evidence suggests that a 20- to 30-minute resting period usually is sufficient to produce the lowest physiological levels (Schneiderman & McCabe, 1989). However, when resting measures are taken in a laboratory setting when the subject is expecting to undergo an experimental manipulation, there is reason to suppose that measured levels might be influenced either by anticipatory arousal and/or by the subject's reactions to the novelty of the experimental setting. In this regard, Obrist (1981) contended that "true" baseline values may be recorded only at a subsequent and separate relaxation session, when the subject is given no experimental tasks to perform. Kamarck, Jennings and Manuck (in 1993) have also developed procedures for reducing variability of baseline measurements by standardizing the subject's activities and thoughts during a resting baseline period by introducing a minimally demanding task during this period.

Other Noninvasive Measurement Techniques

Because this chapter is written at an introductory level, more sophisticated noninvasive cardiac measures that require considerable expertise to implement were not reviewed here. In this regard, procedures to indirectly assess cardiac output, cardiac contractility, and other indices of left ventricular function as well as parasympathetic influences on the heart have been developed and are being increasingly utilized in stress research. The reader is referred to Coles et al. (1986), and Schneiderman et al. (1989), Wilson et al. (1989), for a consideration of these procedures. One technique that is becoming increasingly popular with psychophysiologists, is impedance cardiography (see Wilson et al., 1989 for details). This technique provides information that enables the estimation of cardiac output, stroke volume, and cardiac contractile function. The measurement of these various components of the cardiac responses during stress allows the researcher to make inferences about the underlying neural mechanisms that lead to particular stress-related changes.

Conclusions

This chapter has provided an overview of issues and methods involved in the utilization of cardiovascular measures as physiological measures of the stress response. We have emphasized the importance of recognizing the homeostatic role of the cardiovascular system, as well as the connections among the various parts of the cardiovascular system such that changes in one component of the cardiovascular system necessarily affect other components of the system (Forsyth, 1974). In addition, it is important to emphasize the relationship of cardiovascular responses to behavioral task demands, and the lack of a generalized and nonspecific physiological response to stress. Awareness of the physiology of the cardiovascular system and the psychophysiology of human response to environmental stressors will contribute to the informed use of cardiovascular measures and their appropriate interpretation in stress research.

Acknowledgment

Preparation of this chapter was supported by a grant from the National Institutes of Health (HL47337). The opinions and assertions expressed herein are those of the authors and should not be construed as reflecting those of the USUHS or the Department of Defense.

References

Ax, A. F. (1953). The physiological differentiation between fear and anger in humans. *Psychosomatic Medicine, 15*, 433–442.

Ayman, P., & Goldshine, A. D. (1940). Blood pressure determinations by patients with essential hypertension. I. The difference between clinic and home readings before treatment. *American Journal of Medical Science, 200*, 465–474.

Bigger, T. J., Fleiss, J. L., Rolnitzky, L. M., & Steinman, R. C. (1993). The ability of several short-term measures of RR variability to predict mortality after myocardial infarction. *Circulation, 88*, 927–934.

Brod, J., Fencl, V., Heil, Z., & Jirka, J. (1959). Circulatory changes underlying blood pressure elevation during acute emotional stress in normotensive and hypertensive subjects. *Clinical Science, 18*, 269–279.

Cacioppo, J. T., & Tassinary, L. G. (1990). *Principles of psychophysiology: Physical, social, and inferential elements.* New York: Cambridge University Press.

Coles, M.G.H., Donchin, E., & Porges, S. W. (1986). *Psychophysiology: Systems, processes, and applications.* New York: Guilford Press.

Comstock, G. W. (1957). An epidemiologic study of blood pressure levels in a biracial community in the Southern United States. *American Journal of Hygiene, 65*, 273.

Dunbar, F. (1947). *Mind and body: Psychosomatic medicine.* New York: Random House.

Ekman, P., Levenson, R. W., & Friesen, W. V. (1983). Nervous system activity distinguishes among emotions. *Science, 221*, 1208–1210.

Engel, B. T. (1972). Response specificity. In N. S. Greenfield, & R. A. Sternbach (Eds.), *Handbook of psychophysiology* (pp. 571–576). New York: Holt, Rinehart & Winston.

Forsyth, R. P. (1974). Mechanisms of the cardiovascular responses to environmental stressors. In P. A. Orbist, A. H. Black, J. Brener, & L. V. DiCara (Eds.), *Cardiovascular psychophysiology* (pp. 5–32). Chicago: Aldine Publishing.

Gunn, L. G., Wolf, S., Block, R. T., & Person, R. J. (1972). Psychophysiology of the cardiovascular system. In N. S. Greenfield, & R. A. Sternbach (Eds.), *Handbook of psychophysiology* (pp. 457–483). New York: Holt, Rinehart & Winston.

Harshfield, G. A., Hwang, C., Blank, S. G., & Pickering, T. G. (1989). Research techniques for ambulatory blood pressure monitoring. In N. Schneiderman, S. M. Weiss, & P. G. Kaufmann (Eds.), *Handbook of research methods in cardiovascular behavioral medicine* (pp. 293–310). New York: Plenum Press.

Hassett, J. (1978). *A primer of psychophysiology.* San Francisco: W. H. Freeman.

Heller, L. J., & Mohrman, D. E. (1981). *Cardiovascular physiology.* New York: McGraw-Hill.

Jiang, W., Hayano, J., Coleman, E. R., Hanson, M. W., Frid, D. J., O'Cononor, C., Thurber, D., Waugh, R. A., & Blumenthal, J. A. (1993). Relation of cardiovascular responses to mental stress and cardiac vagal activity in coronary artery disease. *American Journal of Cardiology, 72,* 551–554.

Kligfield, P. (1989). Ambulatory electrocardiographic monitoring: Methods and applications. In N. Schneiderman, S. M. Weiss, & P. G. Kaufmann (Eds.), *Handbook of research methods in cardiovascular behavioral medicine* (pp. 273–292). New York: Plenum Press.

Kamarck, T. W., Jennings, J. R., & Manuck, S. B. (1993). Psychometric applications in the assessment of cardiovascular reactivity. *Homeostasis in health and disease, 34,* 229–243.

Krantz, D. S., Baum, A., & Singer, J. E. (1983). *Handbook of psychology and health (Vol. 3): Cardiovascular disorders and behavior.* Hillsdale, NJ: Laurence Erlbaum.

Krantz, D. S., Gabbay, F. H., Hedges, S. M., Leach, S. G., Gottdiener, J. S., & Rozanski, A. (1993). Mental and physical triggers of silent myocardial ischemia: Ambulatory studies using self-monitoring diary methodology. *Annals of Behavioral Medicine, 15,* 33–40.

Krantz, D. S., & Manuck, S. B. (1984). Psychophysiologic reactivity and risk of cardiovascular disease: A review and methodological critique. *Psychological Bulletin, 96,* 435–464.

Krantz, D. S., Manuck, S. B., & Wing, R. R. (1987). Psychological stress and task variables as elicitors of reaction. In K. A. Matthews, S. M. Weiss, T. Detre, T. M. Dembroski, B. Falkner, S. B. Manuck, & R. B. Williams (Eds.), *Handbook of stress, reactivity, and cardiovascular disease* (pp. 85–108). New York: John Wiley.

Krantz, D. S., & Ratliff-Crain, J. (1989). The social context of stress and behavioral medicine research: Instruction, experimenter effects, and social interactions. In N. Schneiderman, S. M. Weiss, & P. G. Kaufmann (Eds.), *Handbook of research methods in cardiovascular behavioral medicine* (pp. 383–392). New York: Plenum Press.

Lacey, J. I. (1967). Somatic response patterning and stress: Some revisions of activation theory. In M. H. Appley & R. Trumble (Eds.), *Psychological stress* (p. 14). New York: Appleton-Century-Crofts.

Larsen, P. B., Schneiderman, N., & Pasin, R. D. (1986). In M.G.H. Coles, E. Donchin, & S. W. Porges (Eds.), *Psychophysiology: Systems, processes, and applications* (pp. 122–165). New York: Guilford Press.

Manuck, S. B., Kasprowicz, A. L., Monroe, S. M., Larkin, K. T., & Kaplan, J. R. (1989). Psychophysiologic reactivity as a dimension of individual differences. In N. Schneiderman, S. M. Weiss, & P. G. Kaufmann (Eds.), *Handbook of research methods in cardiovascular behavioral medicine* (pp. 365–382). New York: Plenum Press.

Matthews, K. A., Weiss, S. M., Detre, T., Dembroski, T. M., Falkner, B., Manuck, S. B., & Williams, R. B. (1987). *Handbook of stress, reactivity, and cardiovascular disease.* New York: John Wiley.

Obrist, P. A. (1981). *Cardiovascular physiology: A perspective.* New York: Plenum Press.

Papillo, J. F., & Shapiro, D. (1990). The cardiovascular system. In J. T. Cacioppo & L. G. Tassinary (Eds.), *Principles of psychophysiology: Physical, social, and inferential elements* (pp. 456–512). New York: Cambridge University Press.

Patterson, S. W., Piper, H., & Starling, E. H. (1914). The regulation of the heart beat. *Journal of Physiology (London), 48,* 465–513.

Pickering, G. (1968). *High blood pressure* (2nd edition) New York: Grune & Stratton.

Pickering, T. G. (1993). Applications of ambulatory blood pressure monitoring in behavioral medicine. *Annals of Behavioral Medicine, 15,* 26–32.

Pickering, T. G., & Blank, S. G. (1989). The measurement of blood pressure. In N. Schneiderman, S. M. Weiss, & P. G. Kaufmann (Eds.), *Handbook of research methods in cardiovascular behavioral medicine* (pp. 69–80). New York: Plenum Press.

Pickering, T. G., Cvetkovski, B., & James, G. D. (1986). An evaluation of electronic recorders for self-monitoring of blood pressure. *Journal of Hypertension, 4* (Suppl.), S329–S330.

Pickering, T. G., Harshfield, G. A., Kleinert, H. D., Blank, S., & Laragh, J. H. (1982). Blood pressure during normal daily activities, sleep, and exercise: Comparison of values in normal and hypertensive individuals. *Journal of the American Medical Association, 247,* 992–996.

Rushmer, R. F. (1976). *Cardiovascular dynamics* (4th edition). Philadelphia: W. B. Saunders.

Schneiderman, N., Ironson, G. H., & McCabe, P. M. (1987). Physiology of behavior and blood pressure regulation in humans. In S. Julius, & D. R. Bassett (Eds.), *Handbook of Hypertension* (pp. 19–42). New York: Elsevier.

Schneiderman, N., & McCabe, P. M. (1989). Psychophysiologic strategies in laboratory research. In N. Schneiderman, S. M. Weiss, & P. G. Kaufmann (Eds.), *Handbook of research methods in cardiovascular behavioral medicine* (pp. 349–364). New York: Plenum Press.

Schneiderman, N., & Pickering, T. G. (1987). Cardiovascular measures of psychologic reactivity. In K. A. Matthews, S. M. Weiss, T. Detre, T. M. Dembroski, B. Falkner, S. B. Manuck, & R. B. Williams (Eds.), *Handbook of stress, reactivity, and cardiovascular disease* (pp. 145–186). New York: John Wiley.

Schneiderman, N., Weiss, S. M., & Kaufmann, P. G. (1989). *Handbook of research methods in cardiovascular behavioral medicine.* New York: Plenum Press.

Schwartz, G. E., Weinberger, D. A., &Singer, J. A. (1981). Cardiovascular differentiation of happiness, sadness, anger, and fear following imagery and exercise. *Psychosomatic Medicine, 43,* 343–364.

Weber, M. A., & Drayer, J. I. M., (1984). *Ambulatory blood pressure monitoring.* New York: Springer-Verlag.

Weiner, H. (1977). *Psychobiology and human disease.* New York: Elsevier.

Weiner, H. (1980). Norepinephrine, epinephrine, and the sympathomimetic amines. In A. G. Gilman, L. S. Goodman, T. W. Rall, & F. Murad (Eds.), *Goodman and Gilman's the pharmacological basis of therapeutics* (pp. 130–144). New York: Macmillan.

Williams (1987). Patterns of reactivity and stress. In K. A. Matthews, S. M. Weiss, T. Detre, T. M. Dembroski, B. Falkner, S. B. Manuck, & R. B. Williams (Eds.), *Handbook of stress, reactivity, and cardiovascular disease* (pp. 109–126). New York: John Wiley.

Willis, R. (1965). *The Works of William Harvey, M.D.: Translated From Latin With a Life of the Author.* New York: Johnson Reprint. (Originally published in London by Sydenham Society, 1847).

Wilson, M. F., Lovallo, W. R., & Pincomb, G. A. (1989). Noninvasive measurement of cardiac functions. In N. Schneiderman, S. M. Weiss, & P. G. Kaufmann (Eds.), *Handbook of research methods in cardiovascular behavioral medicine* (pp. 23–50). New York: Plenum Press.

10

Measurement of Immune Response

Janice K. Kiecolt-Glaser and Ronald Glaser

Preliminary Considerations

The field of psychoneuroimmunology (PNI) has grown very rapidly in the last decade; a number of studies have shown immunological alterations in response to commonplace stressful events such as academic examinations (Glaser et al., 1990), as well as transient laboratory stressors such as mental arithmetic (Kiecolt-Glaser, Cacioppo, Malarkey, & Glaser, 1992). In addition, although data are limited, chronic stressors have been linked to the longer-term down-regulation of immune function (Baum, Cohen, & Hall, 1993; Kiecolt-Glaser, Dura, Speicher, Trask, & Glaser, 1991), and diverse interventions appear capable of modulating various aspects of immune function (Kiecolt-Glaser & Glaser, 1992). Although the evidence is still preliminary, these immunological changes appear to have consequences for health (Cohen, Tyrrell, & Smith, 1991; Glaser et al., 1987; Kasl, Evans, & Niederman, 1979).

In the first part of this chapter, we provide a very basic overview of some of the important concepts and terms (see also Table 10.1). For more detailed information on behavioral influences on immune function, the reader can consult several sources (Ader, Felten, & Cohen, 1991; Andersen, Kiecolt-Glaser, & Glaser, 1994; Glaser & Kiecolt-Glaser, 1994a, 1994b; Herbert & Cohen, 1993a, 1993b; Kiecolt-Glaser & Glaser, 1992). For further discussions of basic immunology, the November 25, 1992 issue of *JAMA* (Volume 268) provides a primer on allergic and immunological diseases, with 26 articles covering clinically relevant immunology; in addition, there are a number of immunology textbooks (e.g., Male, Champion, Cooke, & Owen, 1991; Stites & Terr, 1991).

Basic Immunological Terms and Concepts

The immune system is the body's defense against infectious and malignant disease. Its two primary tasks are the identification of "self" from "nonself," and the destruction, inactivation, or elimination of foreign substances or materials (nonself). An *antigen* is any molecule identified that can trigger an immune response; for

Table 10.1. Glossary of Important Concepts and Terms

Antibodies	Immunoglobulins. In humans, there are five major classes of antibodies—IgG, IgA, IgM, IgE, and IgD. Some antibodies can neutralize the effects of toxins; others can lyse cell membranes; whereas the IgE class is involved in hypersensitivity reactions.
Blastogenesis	Term used to describe the induction of cell division after exposure to a mitogen or an antigen. Blastogenesis is thought to provide an in vitro model of the lymphocyte proliferative response to challenge by infectious agents.
B-lymphocytes	Lymphocytes that produce immunoglobulins, and the primary cells associated with the humoral immune system; derived from bone marrow.
Cellular immune response	Immune functions not involving antibody but involving primarily T-lymphocytes. Cellular immunity is particularly important for the defense against intracellular viruses, transplanted tissue, cancer cells, fungi, and protozoans.
Helper T-lymphocytes	Cells that stimulate the production of immunoglobulins by B-lymphocytes.
Humoral immune response	The arm of the immune response responsible for the production of immunoglobins.
Hypersensitivity	Allergy, or the enhanced responsiveness of the immune system to a foreign substance which leads to pathological tissue changes. Immediate hypersensitivity reactions can occur in minutes; well-known examples are hayfever, asthma, and hives.
Immunoglobulins	Antibodies
In vitro	Measured under artificial conditions in the laboratory—e.g., in a petri dish or test tube; as opposed to in vivo, in the body.
Leukocytes	White blook cells. See lymphocytes.
Lymphocytes	The majority of leukocytes are lymphocytes, white blood cells that are important for making antibody, as well as specifically taking part in surveying for and eliminating tumor cells, and cells carrying infectious agents.
Lymphokines	Cell products that serve as chemical mediators of lymphocyte functions.
Mitogens	Substances that induce lymphocyte proliferation.
Natural killer (NK) cells	Cells that are thought to provide an important defense against cancer and virus-infected cells.
Suppressor T-lymphocytes	Act to shut off helper T-lymphocytes when sufficient antibody has been produced.
T-lymphocytes	Thymus-derived lymphocytes, critical to the functioning of the cellular immune response.

example, a virus, a bacterium, portions or products of viruses or bacteria, and allergens can act as antigens.

The organs of the *immune system* include the thymus, bone marrow, lymph nodes, spleen, tonsils and adenoids, and Peyer's patches (the latter are located in the small intestine). Collectively, these "lymphoid" organs play roles in the growth, development, and deployment of lymphocytes (white blood cells).

The two major arms of the immune system are the *humoral* immune system and the *cellular* immune system. In the former, *B-lymphocytes* produce *antibodies* or *immunoglobulins*—serum proteins that are induced by and react with antigens with exquisite specificity. The humoral immune response is important for defense against bacteria and viruses in body fluids.

The *cellular immune response,* the non–antibody-producing arm of the immune system, is important for defense against intracellular viruses, transplanted tissue, cancer cells, fungi, and protozoans; it has cells that move around the body that can kill such target cells. *T-lymphocytes* (thymus-derived lymphocytes), the lympho-cytes crucial for cellular immune system function, have a number of subgroups. For example, *cytotoxic T-lymphocytes* migrate to the invasion site in the body, attach themselves to cells expressing foreign antigens, and produce cytotoxic factors that destroy the cells.

Two T-lymphocyte subpopulations have particular importance because of their regulatory effects on immunity. *Helper T-cells* stimulate B-lymphocytes to produce antibody, whereas *suppressor T-lymphocytes* act to shut off helper T-cells when sufficient antibody has been produced. Significant disturbances in the helper/suppressor cell ratio can have important health consequences. Low ratios are found in patients with immunodeficiency disorders. In contrast, high helper/suppressor cell ratios are found in some naturally occurring autoimmune diseases—for example, systemic lupus erthyematosus, hemolytic anemia, severe atopic eczema, and inflammatory bowel disease; in these cases, the immune system appears unable to discriminate self from nonself, and attacks the body's own cells. The loss of suppressor cells may occur for a number of reasons, including their destruction by autoantibodies (antibodies made against self).

Lymphocytes and other kinds of blood cells can synthesize *lymphokines* (also called *cytokines)*—essential chemical mediators of various aspects of the immune response. The *interferons* (IFNs) and the *interleukins* (ILs) are two broad groups of cytokines.

Another component of the cellular immune response, *natural killer* (NK) cells, serve a vital immunological function: They defend against cancer and virus-infected cells. NK cells form an antitumor surveillance system, and appear to be critical in prevention of tumor growth and metastases as has been shown in animal models (Whiteside, Bryant, Day, & Herberman, 1990). They are labeled "natural" killers because they do not require prior exposure to a particular antigen and can therefore kill target cells in a nonspecific manner.

The central nervous system (CNS) and the immune system can communicate through multiple pathways. *Hormones* are very responsive to certain emotional states, and hormones can mediate immune function (e.g., see Baum & Grunberg, Chapter 8, this volume; Ader et al., 1991; Malarkey, Kiecolt-Glaser, Pearl, & Glaser, 1994). In addition, there may be direct connections as well; for example, nerve terminals have been found in the spleens of rats that are in physical contact with T lymphocytes (Ackerman, Bellinger, Felten, & Felten, 1991).

These CNS–immune system interactions can produce rapid changes, as shown in a recent series of studies of acute laboratory stressors (using human subjects) that generally last a half hour or less (reviewed in Kiecolt-Glaser et al., 1992). In contrast to the decrements in lymphocyte numbers reported in some studies of longer-term naturalistic stressors (Herbert & Cohen, 1993b), acute laboratory stressors appear to increase cell numbers in some lymphocyte subpopulations. One possible mechanism may be the acute secretion of stress-responsive hormones, particularly *catecholamines,* which can alter a number of aspects of immune function (Rabin et al., 1989).

In fact, the immunological changes observed following short-term stressors are very similar to those that have been described following epinephrine injections (Crary et al., 1983a, 1983b). These epinephrine-induced changes are thought to reflect transient alterations in lymphocyte migration from lymphoid organs and peripheral blood mediated through receptors on lymphocytes or via the sympathetic nervous system innervation of lymphoid organs like the spleen (Ackerman et al., 1991; Crary et al., 1983a, 1983b; Ottaway & Husband, 1992); patients whose spleens have been removed show much smaller changes in response to an epinephrine infusion than do normal subjects (Van Tits et al., 1990). These latter studies are critical to the interpretation of immunological data from acute laboratory stressors; that is, transient alterations in lymphocyte subpopulations are thought to reflect simple changes in the distribution of cells in circulation in peripheral blood (a process called "trafficking"), not a real change in cell numbers.

Immunological Assays: Basic Information

In order to measure different aspects of immune function, the numbers and/or functional abilities of subgroups of leukocytes (white blood cells) are assayed in blood samples. Immunological assays can be roughly divided into two categories. *Functional assays* reflect the "performance" or the functional efficacy of the cells. In contrast, *enumerative assays* provide information on percentages or numbers of cells. Cell numbers and cell function are not necessarily correlated (e.g., cells may not be differentiated or activated).

A number of leukocyte subpopulations perform specialized immunological functions, and no single immunological assay provides a global measure of immune system function; for this reason, PNI studies typically include a battery of assays. However, because of the interdependence of the various components of the immune system, adverse changes in one subpopulation of lymphocytes, for example, may produce multiple, cascading effects.

There are numerous ways to measure various aspects of immune function, and the number of assays is steadily increasing as technology develops. In this section, we limit our discussion to some of the most common assays used in human PNI studies; the references listed in the introduction provide much more extensive information on the range of assays and their interpretation.

Enumerative Assays

Quantification of lymphocyte subpopulations most commonly involves the use of monoclonal antibodies that are directed at specific lymphocyte surface antigens; the monoclonal antibody is coupled to a fluorescent dye. After lymphocytes are incubated with the monoclonal antibody, the percentage of cells that fluoresce can be counted by an instrument called a flow cytometer. For example, if the CD4 monoclonal antibody has been used, then the *percentage* of helper/inducer T-cells within a sample can be assessed. In addition, an absolute lymphocyte count, available as part of a complete blood count (CBC), is needed to convert the percentage of CD4 cells into the *absolute number* of CD4 cells that represent one prognosticator in human immunodeficiency virus (HIV) infection (see Lopez, Fleisher, & deShazo, 1992).

Collection of concurrent CBC data should be routine when lymphocyte percentages are being assessed; data from two meta-analyses suggest that cell numbers may be more strongly related to stressor appraisal or depression than are cell percentages (Herbert & Cohen, 1993a, 1993b). (It should be noted that CBC data generally do not, by themselves, provide useful data for PNI studies since simple lymphocyte counts are not particularly informative in "normal" populations.)

Functional Assays

In our experience, functional assays are more strongly and reliably related to psychological stressors than are enumerative assays (e.g., Kiecolt-Glaser & Glaser, 1991). In addition, functional assays are essential when one is studying older adult populations, as will be discussed shortly.

Blastogenesis. Lymphocytes are normally found in a resting, nonreplicative state. In order to react to an infection and induce protection, cells need to be activated to replicate and to produce high levels of cytokines. When the immune system has identified and processed an antigen, both T- and B-lymphocytes are induced to proliferate and differentiate into functional subpopulations. A given antigen will stimulate this sequence for only the small subset of cells that have specific compatible receptors; all lymphocytes carry surface receptors that recognize one specific antigen, like a lock and key.

However, the in vitro use of mitogens (substances used in the laboratory that have the ability to stimulate lymphocyte proliferation or replication for large subsets of lymphocytes; analogous to a master key) can provide information on the immune system's ability to respond to certain foreign substances. The proliferative response of both T- and B-lymphocytes to stimulation by *mitogens* (termed *blastogenesis*) such as phytohemagglutinin (PHA, which stimulates T-cell proliferation), pokeweed mitogen (PWM, which stimulates both T- and B-cells) and concanavalin A (Con A, another T-cell mitogen) is thought to provide a model of the body's response to challenge by infectious agents such as bacteria or viruses (Reinherz & Schlossman, 1980).

Typically, blastogenesis involves incubation of lymphocytes with a mitogen in tissue culture media that includes a radioactive isotope. As lymphocytes replicate, they incorporate the isotope into cellular DNA. Proliferation (cell division) can be quantified by an instrument that measures the emission of radiation expressed in "counts per minute" (cpm), thus providing a measure of radioisotope uptake or utilization as a function of cell division—that is, the response to the mitogen/antigen.

Blastogenesis is one of the few immunological assays that has been reliably associated with relevant health parameters. Decreased lymphocyte proliferation reflects the down-regulation of normal immune responses in a variety of immunodeficiency conditions, including acquired immune deficiency syndrome (AIDS) (Fletcher, Baron, Asman, Fischl, & Klimas, 1987), as well as less severe illnesses (e.g., see Cogen, Stevens, Cohen-Cole, Kirk, & Freeman, 1982; Lumino, Welin, Hirvonen, & Weber, 1983), and even normal aging (Roberts-Thompson, Whittingham, Youngchaiyud, & MacKay, 1974).

Because two common mitogens, Con A and PHA, are inexpensive and easily obtained, they are used routinely to induce T-lymphocyte proliferation. They do this in a nonspecific way that does not involve the T-cell receptor. Mitogens are used to study T-cell responses. However, it is known that *antigens* induce T-cell proliferation by binding (specifically) to the T-cell receptor; when this happens, the T-cell is triggered to undergo cell division. A monoclonal antibody to the T3 receptor provides another way to measure T-cell proliferation. When this monoclonal antibody binds to the T-cell receptor, it, too, induces cell division in a manner similar to that induced by an antigen (e.g., Kiecolt-Glaser et al., 1993).

NK Cell Activity. A number of PNI studies have assessed the ability of NK cells to lyse or destroy "target" cells (usually cells from a tumor cell line), a process referred to as NK lysis (e.g., Kiecolt-Glaser, Garner, Speicher, Penn, & Glaser, 1984; Kiecolt-Glaser et al., 1985, 1993). To measure NK cell activity, target cells are grown in a media that is supplemented with a radioisotope; they incorporate the isotope into the cytoplasm of the cell. When NK cells are subsequently incubated with these specially prepared target cells, they lyse the target cells, releasing the isotope; the efficacy of lysis or killing is determined by measuring the amount of isotope released in the process. NK cell activity has a moderate to large effect size in various "stress" or depression studies (Herbert & Cohen, 1993a, 1993b).

Latent Herpesvirus Antibody Titers. The immune system "remembers" pathogens it has previously met, and immunological memory is an important principle that underlies the success of vaccination programs. For example, once a person has been exposed to an infectious agent (e.g., poliovirus) the immune system mounts a defense and eradicates the invading agent; thereafter, that person is very unlikely to develop a clinical illness associated with that infectious agent again because the immune system can quickly destroy the agent should he or she be reexposed. However, some viruses are capable of hiding in a latent state within specific host cells and thus can escape destruction by the immune system; HIV and the herpesviruses provide notable examples. Assays that reflect alterations in herpesvirus latency appear to be quite sensitive to a variety of stressors (Glaser & Kiecolt-Glaser, 1994a; Kasl et al., 1979; Kiecolt-Glaser et al., 1993; McKinnon, Weisse, Reynolds, Bowles, & Baum, 1989)

Individuals will remain latently infected for life after infection with a herpesvirus. The competence of the cellular immune system is thought to be a critical factor in controlling the primary herpesvirus infection, as well as subsequent control of virus latency (Glaser & Kiecolt-Glaser, 1994a). In compromised cellular immunity (e.g.., in patients with immunosuppressive diseases like AIDS; or in patients undergoing immunosuppressive therapies as in organ transplants, the immune system's control over latent herpesvirus replication is impaired. In these cases, reactivation of latent herpesviruses can occur and may result in disease. Furthermore, there are also characteristic elevations in herpesvirus antibody titers that can occur in the absence of any symptoms (Glaser & Kiecolt-Glaser, 1994a). These elevations in herpesvirus antibody titers are thought to reflect the memory antibody response to increased synthesis of the virus or virus proteins. When the cellular immune system becomes more competent (e.g., after cessation of immunosuppressive therapies), there are normally decrements in herpesvirus antibody titers. Thus, although the conclusion seems counterintuitive, *higher* antibody titers to latent herpesviruses suggest that the cellular immune system is *less competent* in controlling herpesvirus latency (Glaser & Kiecolt-Glaser, 1994a).

Antibody titers to latent herpesviruses show reliable changes in response to psychosocial stressors in asymptomatic individuals, particularly Epstein–Barr virus (EBV) and herpes simplex type 1 (HSV-1) (e.g., Glaser, Kiecolt-Glaser, Speicher, & Holliday, 1985; Glaser et al., 1987, 1991; Kiecolt-Glaser et al., 1985, 1991, 1993); in fact, herpesvirus antibody titers have shown the most consistent relationships to psychosocial variables of any of the diverse immunological assays we have used in our laboratory. In accordance with the suggestion that elevations in herpesvirus antibody titers reflect a broader down-regulation of cellular immune function, we have found also that specific T-cell killing of EBV-infected target cells decreased, synthesis of an induced lymphokine was altered (Glaser et al., 1987), and the proliferative response to several EBV polypeptides (viral proteins) also decreased in association with stress (Glaser et al., 1993).

Other Functional Assays. We have limited our descriptions to some of the most commonly used functional immune assays. Although an extended discussion of less routine assays is not possible in this chapter, investigators have also found stress-related differences in such diverse aspects of immune function as antibody and virus specific T-cell responses to a viral vaccine (Glaser, Kiecolt-Glaser, Bonneau, Malarkey, & Hughes, 1992; Jabaaij et al., 1993); the production of two key lymphokines, gamma interferon and interleukin-2 (IL-2) (Glaser et al., 1987, 1990); alterations in the expression of a receptor for IL-2 on the surface of lymphocytes; and modulation of IL-2 receptor gene expression (Glaser et al., 1990)

Choosing Appropriate Immunological Assays

Investigators interested in the role that various stressors may play in the incidence, severity, or duration of infectious diseases, cancer, immunodeficiency diseases, or autoimmune diseases might be interested in certain aspects of immune function in their subjects. Alternatively, if a researcher simply wants some kind of "biological" marker to supplement psychosocial measures of stressor appraisal without relevance to any of the above diseases, then endocrine or autonomic indices may prove to be easier and cheaper substitutes (see Chapters 8 & 9).

What aspect(s) of immune function should you measure? The answer to the question is determined by the experimental questions of interest, the expertise and interests of one's collaborating immunologist, the populations studied, the amount of blood that can be drawn from each subject, available funds, and other logistical constraints.

Collaborating with an Immunologist

The development of a collaborative relationship with an immunologist is the most critical element for a behavioral scientist who wishes to begin PNI research. The immunologist will help design studies with an eye to the methodological and logistical constraints discussed in this chapter, choose assays that are appropriate to the study population that can be performed in his or her laboratory, review immunological data from the study to ensure both reliability and validity, and provide immunological expertise necessary for the interpretation of results. The behavioral scientist might also form a collaborative relationship with a physician whose clinical specialty involves immunologically mediated disorders (e.g., AIDS or asthma); such a person is likely to be knowledgeable about both clinical issues and associated immunological alterations.

Immunologists represent a diverse group of biological scientists who generally have a relatively narrow focus for their own work and interests. The skills, training, and resources of an immunologist strongly influence his/her choice of assays for joint efforts with a behavioral scientist: for example, cytokine researchers are likely to see cytokine assays as central to understanding alterations in immune function, whereas a herpes virologist/immunologist might argue that assessing aspects of herpesvirus latency provides a window on the competency of the cellular immune response (e.g., Glaser & Kiecolt-Glaser, in 1994a). Thus, the immunologist must be a key player in choosing the type of assay; by way of analogy, a behavioral scientist should not expect a neuropsychologist to have any enthusiasm or expertise for a collaborative venture if the scientist insists on using the Rorschach as the project's primary measure of cognitive function.

Sometimes it is not possible to establish a collaborative relationship with an immunologist. Hospital or commercial laboratories will perform many immunological assays on a fee-for-service basis, albeit generally at much greater cost than with a collaborator. If a collaborative relationship is not possible, an immunological consultant who reviews raw data can provide helpful input on the multiple technical problems that are not obvious to an untrained eye, as well as guidance in the interpretation of results.

Matching Immunological Assays to Specific Research Questions

As mentioned earlier, a wide range of immunological functions appear to be stress-responsive; given the limited knowledge about which immunological alterations may be connected to actual health changes, there are no clear guidelines for assays that may be "essential" beyond those that have clear relevance to the health status of the behavioral scientist's proposed population and research questions. For example, the number of CD4 (helper/inducer) T-lymphocytes is one key marker for HIV progression (Lopez et al., 1992). The function and numbers of NK cells (particularly their function) appear relevant to a number of cancers, particularly the spread of metastatic cancer (Whiteside et al., 1990). The number of suppressor cells and the helper/suppressor ratio provide important information for certain autoimmune diseases (e.g., systemic lupus erthyematosus, hemolytic anemia, severe atopic eczema, and inflammatory bowel disease); in these diseases, the loss of suppressor cells may correlate with clinical severity (Reinherz & Schlossman, 1980). Vaccine studies normally involve the assessment of antibody production and a T-cell and/or cytokine response to vaccine antigens (Glaser et al., 1992; Jabaaij et al., 1993). Researchers who are infecting subjects with a specific virus (e.g., a cold virus) need to measure antibody to specific viral antigens both before and after infection, as well as aspects of the cellular immune response such as the virus-specific T-lymphocyte response.

Age-related immunological decrements are thought to be associated with the greatly increased morbidity and mortality from infectious illness observed in the elderly; for example, mortality from influenza infection is four times greater among people who are over 60, compared to those younger than 40 (Burns, Lum, Seigneuret, Giddings, & Goodwin, 1990). These age-related immunological changes are demonstrated in the functional aspects of the cellular immune response, but only minimally or not at all with respect to cell numbers (Murasko, Weiner, & Kaye, 1988; Wayne, Rhyne, Garry, & Goodwin, 1990). Thus, functional assays are the highest priority when the health of the elderly is of interest.

As mentioned earlier, no single immunological assay provides a global measure of immune system function; thus, researchers should utilize several different assays to assess different aspects of immune function. In addition to choosing assays based on population-specific questions, researchers may also wish to consult two meta-analyses that review the relationships among stress depression, and immune function. Herbert and Cohen (1993a, 1993b) describe the effect sizes for a number of common immunological assays; these articles provide helpful information for choosing more psychosocially "sensitive" assays, as well as data on which to base the number of subjects needed to detect certain effects.

Immunological Data and Health Status: Infectious Illness

It is sometimes erroneously assumed that stress-related alterations in immune function translate directly into changes in health, and that immunological data can serve as a surrogate measure of health status. In fact, the extent to which relatively small immunological changes affect the incidence, severity, or duration of immunologically relevant diseases is unknown. The type, intensity, and chronicity of a stressor; the degree and pervasiveness of immune modulation; and an individual's prior immunological and health status, genetic background, and exposure to an infectious agent are fundamental factors in determining whether actual health changes will occur.

Moreover, there are many problems in clearly demonstrating causal relationships between poorer immune function and illness, particularly infectious illnesses. Infectious illnesses occur relatively infrequently in the general population, with most adults reporting only a few illness episodes a year. As a consequence, alterations in low base rates are difficult to detect, particularly with the relatively small sample sizes necessitated by the time and expense inherent in PNI research. Moreover, exposure to pathogens is essential for development of infection, but exposure is not simply a random event; for example, families with small children are likely to have a higher incidence of illness, whereas socially isolated individuals are less likely to be exposed to pathogens.

Alternatively, in order to demonstrate causal relationships between psychosocial stressors and the development of infectious illness, investigators have inoculated subjects with a pathogen, a vaccine, or a harmless antigen. By evaluating the timing and strength of antibody and T-cell or cytokine response following inoculation, a researcher may model the body's response to infection.

For example, we gave each of a series of three hepatitis B vaccine inoculations to 48 medical students on the last day of three 3-day examination series to study the effect of an academic stressor on the students' ability to generate an immune response to a *primary antigen*—that is, an antigen to which they had no previous exposure (Glaser et al., 1992). A quarter of the students *seroconverted* (produced an antibody response to the vaccine) after the first injection, and they reported feeling less stressed and less anxious than those students who did not seroconvert until after the second injection. In addition, students who reported greater social support demonstrated a stronger immune response to the vaccine at the time of the third inoculation, as measured by antibody titers to one hepatitis B antigen, and the blastogenic response to one of the hepatitis B viral peptides (proteins). These stress-related alterations in hepatitis B vaccine response have subsequently been replicated by another laboratory (Jabaaij et al., 1993). These data suggest that the immunological response to a vaccine can be modulated by a relatively mild stressful event in young, healthy adults.

Vaccine response data such as these can provide a window on the body's response to other pathogens, such as viruses or bacteria; individuals who show a delayed or blunted vaccine response could be at greater risk for more severe illness.

Research with older adults provides further support for these assumptions. Many older adults do not respond to vaccines (or other "new" antigens) as efficiently as younger adults (Phair, Kauffman, Bjornson, Adams, & Linnemann, 1978). Older adults attain lower peak antibody levels after vaccination, and show more rapid or steeper rates of decline than do younger adults in their immune response to influenza and other antigens (Burns et al., 1990). These age-related immunological decrements are thought to be associated with the greatly increased morbidity and mortality from infectious illness in the elderly; for example, among adults over 75 years of age, pneumonia and influenza together are the fourth leading cause of death (Yoshikawa, 1983). Thus, this same age group shows poorer vaccine responses and greater morbidity and mortality from infectious illnesses.

Researchers have also inoculated subjects with a live virus. Cohen, Tyrrell, and Smith (1991) prospectively studied the relationship between stress appraisals and susceptibility to colds by inoculating volunteers with five different cold viruses or a placebo. They found that the rates of both respiratory infection and clinical colds increased in a dose–response manner with increases in stress. Studies such as these in which subjects are inoculated with pathogens or vaccines provide researchers with a means of controlling exposure and concentration of a pathogen; moreover, because immune function may be assessed prior to the infectious challenge, these studies provide better data on causality than is possible with naturally occurring infections.

Intervention Studies

Researchers who wish to try to enhance immune function via an intervention need to consider the initial immunological status of their potential subjects. If an individual's immune system is functioning satisfactorily, it may not be possible to "enhance" immune function above normal levels; in fact, it is possible that it would be undesirable to do so. More is not necessarily better; for example, an overactive immune system may lead to autoimmune disease. In the absence of any age-, disease- or stress-related downward alterations in a study population's immune function, any intervention designed to enhance immune function could fail to alter immune function because of homeostatic regulation; if effective in enhancing immune function, it could be maladaptive.

Changes Following "Laboratory" or Other Brief Stressors

Responses to acute laboratory stressors show considerable variability among individuals and across situations. Individual differences in cardiovascular reactivity have been studied extensively (see Chapter 9 for details; also see Manuck, Kasprowicz, Monroe, Larkin, & Kaplan, 1989); since cardiovascular and catecholaminergic reactivity tend to co-vary when assessed under the same conditions, researchers have analyzed immunological changes in relationship to cardiovascular reactivity (Bachen et al., 1992; Manuck, Cohen, Rabin, Muldoon, & Bachen, 1991; Sgoutas-Emch et al., 1994). High-reactivity subjects demonstrate greater immunological change than low-reactivity subjects, with the latter showing little or no change (reviewed in Kiecolt-Glaser et al., 1992). Both the duration and intensity of psychological stressors (as indexed by cardiovascular changes) are related to the breadth and magnitude of immune changes in laboratory studies. Obviously, at a minimum, heart rate and blood pressure measurements (and, ideally, plasma catecholamines) are needed to aid comparisons across studies with various acute laboratory stressors.

Logistic Issues

In the ideal study, blood samples would be collected from all experimental and control subjects at precisely the same time. The realities of research normally make such a plan impossible. When samples are collected on multiple days from groups of subjects who are hypothesized to differ on some characteristic, blood samples from the different subject cohorts need to be assayed simultaneously (e.g., rather than samples being assayed from depressed subjects on one day and nondepressed control subjects on the next). The day on which samples are gathered and assays performed (i.e., the "measurement occasion") can account for as much as 97 percent of between-group differences, depending on the assay and the laboratory (Schleifer, Keller, Bond, Cohen, & Stein, 1989). In order to avoid systematic bias, it is important to intermingle subjects from various groups when samples are collected. In addition, the immunological data obtained from "control" subjects run on the same day as the targeted population can be used statistically to control for daily variation using analysis of partial variance (Schleifer, Haftan, Cohen, & Keller, 1993).

A related issue is the length of time during which blood samples sit before assays are begun in the laboratory. Sample storage time can have significant effects on some immunological parameters (Fletcher et al., 1987). This becomes a particularly important issue when samples are sent to a commercial laboratory, since laboratories may assay samples immediately on some days, but wait up to 24 hours (or more) on other days, since they try to cluster assays to conserve technician time. The error variance related to differences in storage time is likely to be much greater in commercial laboratories. Unless samples can be frozen (and preparation is required before freezing), shipping samples to a distant laboratory will also introduce these same difficulties with variability in time.

All subjects need to be studied at the same time of day (i.e., within the same 1- to 3-hour window, depending on the parameters of interest) to minimize error variance associated with diurnal variation. A few assays, which have longer half-lives, are quite stable so that diurnal variation is not problematic: In particular, the half-life for immunoglobulins (antibodies) is 6 to 8 days for IgA, 9 to 11 days for IgM, and 21 days for IgG.

Immunological assays require considerable time and technical skill. For example, setting up routine assays using blood samples from several individuals generally takes an experienced technician the better part of a day; most of the common assays described in this chapter require fresh cells, which means that blood samples cannot be drawn and then frozen for later processing. Thus, if blood samples arrive at an immunology laboratory in the afternoon, the technician will then need to work through the evening or later. Within limits, however, the time per subject is not additive, and it may not require substantially more time to prepare several samples for the same assay than a single sample. In addition to the savings in technical time, the grouping or clustering of samples is highly desirable because of the "measurement occasion" issues described above.

In order to obtain a sufficient quantity of blood, samples are drawn from the arm; depending on the battery of assays in a particular study, we generally draw 30 to 60 cc (1 to 2 ounces). Although 30 to 60 cc is a small fraction of an adult's total blood volume, it means that a phlebotomist or nurse will need to fill three to six 10-cc tubes, a process that disturbs some subjects much more than others. We always mention that blood draws are a key part of the study when recruiting subjects, and we ask potential subjects if they have any needle phobias; attempting to draw blood from a needle phobic presents a number of obvious problems, particularly fainting.

When repeated blood samples are collected over a period of several hours, use of an indwelling catheter avoids the additional distress and pain produced by repeated venipuncture. Adaptation periods of 30 minutes or more are advisable following catheter insertion (Manuck et al., 1989). The amount of blood that can be obtained from subjects sets limits on the kinds and numbers of immunological assays that can be performed. Very few PNI researchers have attempted to study children or adolescents, in part because of the difficulties involved in obtaining blood samples; in addition, they are likely to have a higher incidence of needle phobias.

One assay, secretory IgA (s-IgA), uses saliva, rather than blood. However, salivary flow changes in response to stressors, providing an additional methodological problem. Studies that use s-IgA need to control adequately for flow rate because of the associated methodological problems that may otherwise make interpretation of such data questionable (Herbert & Cohen, 1993b; Stone, Cox, Valdimarsdottir, & Neale, 1987).

PNI research is expensive, and supplies for a small pilot study can easily cost several thousand dollars, aside from the labor needed for the assays. In addition to the materials themselves, there may be other, less obvious costs; for example, radioisotopes, used in a number of assays including blastogenesis and NK cell lysis, incur expensive charges for disposal of isotope waste. Hospital laboratory charges, although including both supplies and labor, may be prohibitive; for example, analysis of a single blood sample for one assay, NK cell lysis, may cost from $200 to $400.

In the beginning of a particular study, one should buy sufficient quantities of laboratory supplies for the entire study if at all possible (e.g., mitogens, fetal bovine serum, media, plasticware, etc). Although this suggestion may seem trivial or commonsensical, it can have enormous consequences for laboratory data in a given study. For example, we found a 10-fold difference in the relative values obtained for gamma interferon using different lots of Con A in two studies in which lymphocytes were stimulated with Con A to produce gamma interferon (Glaser, Rice, Speicher, Stout, & Kiecolt-Glaser, 1986; Glaser et al., 1987). Thus, if an investigator were to buy different batches of mitogen without attempting to examine their relative potencies, one could easily show remarkable artifactual changes over time. For these same reasons, assays need to be conducted within the same laboratory across a study.

A number of immunological assays do not have "normal" values or ranges for comparison purposes, particularly the functional assays. Moreover, functional assays appear to show greater day-to-day variation than enumerative assays (Schleifer et al., 1993). In addition, protocols may vary from one laboratory to another, and differences in the methods used for immunological assays can produce dissimilar data. For example, the length of time that lymphocytes are routinely incubated with a mitogen may be 48 hours in one laboratory, 72 hours in another; longer incubation times are likely to result in higher "counts per minute"—that is, greater uptake of the radioisotope—while not reflecting real differences in the proliferative response. Some functional assays are typically run in triplicate, and the triplicates are then averaged (e.g., blastogenesis, NK cell lysis)—providing one indication of the inherent variability of the assay.

As noted earlier, distressed individuals are more likely to have life-styles that put them at greater risk, including poorer health habits such as a greater propensity for alcohol and drug abuse, poorer sleep, poorer nutrition, less exercise, etc.; and these health behaviors may make immunological data much more variable (see review in Kiecolt-Glaser & Glaser, 1988). For this reason, careful assessment of health-related behavior is essential. By making such assessments a routine part of any protocol, it may be possible statistically to control some of the error variance related to these factors, thus providing a clearer understanding of psychosocial influences on immune function.

Future Directions

Convergent evidence from several laboratories suggests that chronic stressors may enhance differences in sympathetic nervous system (SNS) reactivity, neuropeptide release, and immune function (Fleming, Baum, Davidson, Rectanus, & McArdle, 1987; Irwin et al., 1992; Kiecolt-Glaser et al., 1992; McKinnon et al., 1989). If sympathetic activation is a marker or determinant of immune function, then longitudinal studies that evaluate the relationships among psychosocial stressors, SNS activity and reactivity, stress-related immune and endocrine changes, and longer-term changes in health are needed to determine whether extrapolations from cross-sectional data on acute events to chronic and longitudinal effects are warranted. Through interdisciplinary collaborations, the measurement of multiple biological stress responses should help investigators understand how psychosocial stressors get translated into adverse health changes.

References

Ackerman, K. D., Bellinger, D. L., Felten, S. Y., & Felten, D. L. (1991). Ontogeny and senescence of noradrenergic innervation of the rodent thymus and spleen. In R. Ader, D. L. Felten, & N. Cohen (Eds.), *Psychoneuroimmunology* (2nd edition, pp. 72–114). San Diego: Academic Press.

Ader, R., Felten, D. L., & Cohen, N. (Eds.). (1991). *Psychoneuroimmunology.* New York: Academic Press.

Andersen, B. L., Kiecolt-Glaser, J. K., & Glaser, R. (1994). A biobehavioral model of cancer stress and disease course. *American Psychologist 49,* 389–404.

Baum, A., Cohen L., & Hall, M. (1993). Control and intrusive memories as possible determinants of chronic stress. *Psychosomatic Medicine, 55,* 274–286.

Bachen, E. A., Manuck, S. B., Marsland, A. L., Cohen, S., Malkoff, S. B., Muldoon, M. F., & Rabin, B. S. (1992). Lymphocyte subset and cellular immune responses to a brief experimental stressor. *Psychosomatic Medicine, 54,* 673–679.

Burns, E. A., Lum, L. G., Seigneuret, M. C., Giddings, B. R., & Goodwin, J. S. (1990). Decreased specific antibody synthesis in old adults: Decreased potency of antigen-specific B cells with aging. *Mechanisms of Aging and Development, 53,* 229–241.

Cogen, R. B., Stevens, A. W., Cohen-Cole, S., Kirk, K., & Freeman, A. (1982). Leukocyte function in the etiology of acute necrotizing ulcerative gengivitis. *Journal of Periodontology, 54,* 402–407.

Cohen, S., Tyrrell, D. A., & Smith, A. P. (1991). Psychological stress in humans and susceptibility to the common cold. *New England Journal of Medicine, 325,* 606–612.

Crary, B., Borysenko, M., Sutherland, D. C., & Kutz, I. Borysenko, J. Z., & Benson, H. (1983a). Decrease in mitogen responsiveness of mononuclear cells from peripheral blood after epinephrine administration in humans. *Journal of Immunology, 130,* 694–497.

Crary, B., Hauser, S. L., Borysenko, M., Ilan, K., Hoban, C., Ault, K., Weiner, H., Benson, H. (1983b). Epinephrine-induced changes in the distribution of lymphocyte subsets in peripheral blood of humans. *Journal of Immunology, 131,* 1178–1181.

Fletcher, M. A., Baron, G. C., Asman, M. R., Fischl, M. A., & Klimas, N. G. (1987). Use of whole blood methods in assessment of immune parameters in immunodeficiency states. *Diagnostic Clinical Immunology, 5,* 69–81.

Fleming, I., Baum, A., Davidson, L. M., Rectanus, E., & McArdle, S. (1987). Chronic stress as a factor in physiologic reactivity to challenge. *Health Psychology, 6,* 221–237.

Glaser, R., Kennedy, S., Lafuse, W. P., Bonneau, R. H., Speicher, C. & Kiecolt-Glaser, J. K. (1990). Psychological stress-induced modulation of IL-2 receptor gene expression and IL-2 production in peripheral blood leukocytes. *Archives of General Psychiatry, 47,* 707–712.

Glaser, R. & Kiecolt-Glaser, J. K. (1994a). Stress-associated immune modulation and its implications for reactivation of latent herpesviruses. In R. Glaser & J. Jones (Eds.), *Human herpesvirus infections* (pp. 245–270). New York: Dekker Glaser, R., & Kiecolt-Glaser, J. K. (Eds.).

Glaser, R., & Kiecolt-Glaser, J. K. (Eds.) (1994b). *Handbook of human stress and immunity.* San Diego: Academic Press.

Glaser, R., Kiecolt-Glaser, J. K., Bonneau, R., Malarkey, W., & Hughes, J. (1992). Stress-induced modulation of the immune response to recombinant hepatitis B vaccine. *Psychosomatic Medicine, 54,* 22–29.

Glaser, R., Kiecolt-Glaser, J. K., Speicher, C. E., & Holliday, J. E. (1985). Stress, loneliness, and changes in herpesvirus latency. *Journal of Behavioral Medicine, 8,* 249–260.

Glaser, R., Pearson, G. R., Bonneau, R. H., Esterling, B. A., Atkinson, C., & Kiecolt-Glaser, J. K. (1993). Stress and the specific memory cytotoxic T-cell response to Epstein–Barr virus. *Health Psychology, 12,* 435–442.

Glaser, R., Pearson, G. R., Jones, J. F., Hillhouse, J., Kennedy, S., Mao, H., & Kiecolt-Glaser, J. K. (1991). Stress-related activation of Epstein–Barr virus. *Brain, Behavior & Immunity, 5,* 219–232.

Glaser, R., Rice, J., Sheridan, J., Fertel, R., Stout, J., Speicher, C. E., Pinsky, D., Kotur, M., Post, A., Beck, M., & Kiecolt-Glaser, J. K. (1987). Stress-related immune suppression: Health implications. *Brain, Behavior, and Immunity, 1,* 7–20.

Glaser, R., Rice, J., Speicher, C. E., Stout, J. C., & Kiecolt-Glaser, J. K. (1986). Stress depresses interferon production by leukocytes concomitant with a decrease in natural killer cell activity. *Behavioral Neuroscience, 100,* 675–678.

Herbert, T. B., & Cohen, S. (1993a). Depression and immunity: A meta-analytic review. *Psychological Bulletin, 113,* 472–486.

Herbert, T. B., Cohen, S. (1993b). Stress and immunity in humans: A meta-analytic review. *Psychosomatic Medicine, 55,* 364–379.

Irwin M., Brown M., Patterson, T., Hauger, R., Mascovich, A., Grant, I. (1992). Neuropeptide Y and natural killer activity: Findings in depression and Alzheimer caregiver stress. *FASEB Journal 5:*3100–3107.

Jabaaij, P. M., Grosheide, R. A., Heijtink, R. A., Duivenvoorden, H. J., Ballieux, R. E., & Vingerhoets, A.J.J.M. (1993). Influence of perceived psychological stress and distress of antibody response to low dose rDNA Hepatitis B vaccine. *Journal of Psychosomatic Research, 37,* 361–369.

Kasl, S. V., Evans, A. S., & Niederman, J. C. (1979). Psychosocial risk factors in the development of infectious mononucleosis. *Psychosomatic Medicine, 41,* 445–466.

Kiecolt-Glaser, J. K., Cacioppo, J. T., Malarkey, W. B., & Glaser, R. (1992). Acute psychological stressors and short-term immune changes: What, why, for whom, and to what extent? *Psychosomatic Medicine, 54,* 680–685.

Kiecolt-Glaser, J. K., Dura, J. R., Speicher, C. E., Trask, O. J., & Glaser, R. (1991). Spousal caregivers of dementia victims: Longitudinal changes in immunity and health. *Psychosomatic Medicine, 53,* 345–362.

Kiecolt-Glaser, J. K., Garner, W., Speicher, C. E., Penn, G., & Glaser, R. (1984). Psychosocial modifiers of immunocompetence in medical students. *Psychosomatic Medicine, 46,* 7–14.

Kiecolt-Glaser, J. K., & Glaser, R. (1988). Methodological issues in behavioral immunology research with humans. *Brain, Behavior, and Immunity, 2,* 67–78.

Kiecolt-Glaser J. K., & Glaser, R. (1991). Stress and immune function in humans. In R. Ader, D. Felten, & N. Cohen (Eds.), *Psychoneuroimmunology* (pp. 849–867). San Diego: Academic Press.

Kiecolt-Glaser, J. K., & Glaser R. (1992). Psychoneuroimmunology: Can psychological interventions modulate immunity? *Journal of Consulting and Clinical Psychology, 60,* 569–575.

Kiecolt-Glaser, J. K., Glaser, R., Strain, E., Stout, J., Tarr, K., Holliday, J., & Speicher, C. (1986). Modulation of cellular immunity in medical students. *Journal of Behavioral Medicine, 9,* 5–21.

Kiecolt-Glaser, J. K., Glaser, R., Williger, D., Stout, J., Messick, G., Sheppard, S., Ricker, D., Romisher, S. C., Briner, W., Bonnell, G., & Donnerberg, R. (1985). Psychosocial enhancement of immunocompetence in a geriatric population. *Health Psychology, 4,* 25–41.

Kiecolt-Glaser, J. K., Malarkey, W. B., Chee, M., Newton, T., Cacioppo, J. T., Mao, H. Y., & Glaser, R. (1993). Negative behavior during marital conflict is associated with immunological down-regulation. *Psychosomatic Medicine, 55,* 395–409.

Lopez, M., Fleisher, T., & deShazo, R. D. (1992). Use and interpretation of diagnostic immunologic laboratory tests. *JAMA, 268,* 2970–2990.

Lumino, J., Welin, M. G., Hirvonen, P., & Weber, T. (1983). Lymphocyte subpopulations and reactivity during and after infectious mononucleosis. *Medical Biology, 61,* 208–213.

Malarkey, W. B., Kiecolt-Glaser, J. K., Pearl, D., & Glaser, R. (1994). Hostile behavior during marital conflict alters pituitary and adrenal hormones. *Psychosomatic Medicine, 56,* 41–51.

Male, D., Champion, B., Cooke, A., & Owen, M. (1991). *Advanced immunology.* Philadelphia: J. B. Lippincott.

Manuck, S. B., Cohen, S., Rabin, B. S., Muldoon, M. F., & Bachen, E. A. (1991). Individual differences in cellular immune response to stress. *Psychology of Science, 2,* 111–115.

Manuck, S. B., Kasprowicz, A. L., Monroe, S. M., Larkin, K. T., Kaplan, J. R. (1989). Psychophysiologic reactivity as a dimension of individual differences. In N. Schneiderman, S. N. Weiss, & P. G. Kaufman (Eds.), *Handbook of research methods in cardiovascular behavioral medicine* (pp. 366–382). New York: Plenum.

McKinnon, W., Weisse, C. S., Reynolds, C. P., Bowles, C. A., & Baum, A. (1989). Chronic stress, leukocyte subpopulations, and humoral response to latent viruses *Health Psychology, 8,* 399–402.

Murasko, D. M., Weiner, P., & Kaye, D. (1988). Association of lack of mitogen-induced lymphocyte proliferation with increased mortality in the elderly. *Aging: Immunology and Infectious Disease, 1,* 1–6.

Ottaway, C. A., & Husband, A. J. (1992). Central nervous system influences on lymphocyte migration. *Brain Behavior and Immunity, 6,* 97–116.

Phair, J., Kauffman, C. A., Bjornson, A., Adams, L., & Linnemann, C. (1978). Failure to respond to influenza vaccine in the aged: Correlation with B-cell number and function. *Journal of Laboratory Clinical Medicine, 92,* 822–828.

Rabin, B. S., Cohen, S., Ganguli, R., Lysle, D. T., & Cunnick, J. E. (1989). Bidirectional interaction between the central nervous system and the immune system. *Critical Reviews in Immunology, 9,* 279–312.

Reinherz, E. L., & Schlossman, S. F. (1980). Current concepts in immunology: Regulation of the immune response—inducer and suppressor T-lymphocyte subsets in human beings. *New England Journal of Medicine, 303,* 370–373.

Roberts-Thompson, I. C., Whittingham, S., Youngchaiyud, U., & MacKay, I. R. (1974). Aging, immune response, and mortality. *Lancet, ii,* 368–370.

Schleifer, S. J., Haftan, M. E., Cohen, J. & Keller, S. E. (1993). Analysis of partial variance (APV) as a statistical approach to control day to day variation in immune assays. *Brain, Behavior, and Immunity, 7,* 243–252.

Schleifer, S. J., Keller, S. E., Bond, R. N., Cohen, J., & Stein, M. (1989). Depression and immunity: Role of age, sex, and severity. *Archives of General Psychiatry, 46,* 81–87.

Sgoutas-Emch, S. A., Cacioppo, J. T., Uchino, B., Malarkey, W., Pearl D., Kiecolt-Glaser, J. K., & Glaser R. (1994). The effects of an acute psychological stressor on cardiovascular, endocrine, and cellular immune response: A prospective study of individuals high and low in heart reactivity. *Psychophysiology, 31,* 264–271.

Stites, D. P., & Terr, A. I. (1991). *Basic and clinical immunology.* San Mateo, CA: Appleton & Lange.

Stone, A. A., Cox, D. S., Valdimarsdottir, H., & Neale, J. (1987). Secretory IgA as a measure of immunocompetence. *Journal of Human Stress, 13,* 136–140.

Van Tits, L. J. H., Michel, M. C., Grosse-Wilde, H., Happell, M., Eigler, F. W., Soliman, A., Brodde (1990). Catecholamines increase lymphocyte β_2-adrenergic receptors via a β_2-adrenergic, spleen-dependent process. *American Journal of Physiology, 258,* E191–E201.

Wayne, S. J., Rhyne, R. L., Garry, P. J., & Goodwin, J. S. (1990). Cell mediated immunity as a predictor of morbidity and mortality in subjects over sixty. *Journal of Gerontology, Medical Sciences, 45,* M45–M48.

Whiteside, T. L., Bryant, J., Day, R., & Herberman, R. B. (1990). Natural killer cytotoxicity in the diagnosis of immune dysfunction: Criteria for a reproducible assay. *Journal of Clinical Laboratory Analysis, 4,* 102–114.

Yoshikawa, T. T. (1983). Geriatric infectious diseases: An emerging problem. *Journal of the American Geriatrics Society, 31,* 34–39.

Subject Index